SEMICONDUCTOR POWER ELECTRONICS

SEMICONDUCTOR POWER ELECTRONICS

Richard G. Hoft
Professor of Electrical Engineering
University of Missouri
Columbia, Missouri

Van Nostrand Reinhold Electrical/Computer Science and Engineering Series

 VAN NOSTRAND REINHOLD COMPANY
———————————————————————————— New York

Published by Van Nostrand Reinhold Company Inc.
115 Fifth Avenue
New York, New York 10003

Van Nostrand Reinhold Company Limited
Molly Millars Lane
Wokingham, Berkshire RG11 2PY, England

Van Nostrand Reinhold
480 Latrobe Street
Melbourne, Victoria 3000, Australia

Macmillan of Canada
Division of Gage Publishing Limited
164 Commander Boulevard
Agincourt, Ontario M1S 3C7, Canada

15 14 13 12 11 10 9 8 7 6 5 4 3 2 1

Library of Congress Cataloging-in-Publication Data

Hoft, R. G. (Richard Gibson)
  Semiconductor power electronics.

  (Van Nostrand Reinhold electrical/computer science and engineering series)

  Bibliography: p.
  Includes index.
  1. Power semiconductors. I. Title.
TK7871.85.H65   1986   621.3815'2   85-20222
ISBN 0-442-22543-1

Van Nostrand Reinhold
Electrical/Computer Science and Engineering Series
Sanjit Mitra — Series Editor

HANDBOOK OF ELECTRONIC DESIGN AND ANALYSIS PROCEDURES USING PROGRAMMABLE CALCULATORS, by Bruce K. Murdock

COMPILER DESIGN AND CONSTRUCTION, by Arthur B. Pyster

SINUSOIDAL ANALYSIS AND MODELING OF WEAKLY NONLINEAR CIRCUITS, by Donald D. Weiner and John F. Spina

APPLIED MULTIDIMENSIONAL SYSTEMS THEORY, by N. K. Bose

MICROWAVE SEMICONDUCTOR ENGINEERING, by Joseph F. White

INTRODUCTION TO QUARTZ CRYSTAL UNIT DESIGN, by Virgil E. Bottom

DIGITAL IMAGE PROCESSING, by William B. Green

SOFTWARE TESTING TECHNIQUES, by Boris Beizer

LIGHT TRANSMISSION OPTICS, Second edition, by Dietrich Marcuse

REAL TIME COMPUTING, edited by Duncan Mellichamp

HARDWARE AND SOFTWARE CONCEPTS IN VLSI, edited by Guy Rabbat

MODELING AND IDENTIFICATION OF DYNAMIC SYSTEMS, by N. K. Sinha and B. Kuszta

COMPUTER METHODS FOR CIRCUIT ANALYSIS AND DESIGN, by Jiri Vlach and Kishore Singhal

HANDBOOK OF SOFTWARE ENGINEERING, edited by C. R. Vick and C.V. Ramamoorthy

SWITCHED CAPACITOR CIRCUITS, by Phillip E. Allen and Edgar Sanchez-Sinencio

SOFTWARE SYSTEM TESTING AND QUALITY ASSURANCE, by Boris Beizer

MODERN DC-TO-DC SWITCHMODE POWER CONVERTER CIRCUITS, by Rudolf P. Severns and Gordon E. Bloom

ESTIMATION AND OPTIMUM CONTROL OF SYSTEMS, by Theodore F. Elbert

DIGITAL TELEPHONY, by Bernhard Keiser and Eugene Strange

PREFACE

Semiconductors have been used widely in signal-level or "brain" applications. Since their invention in 1948, transistors have revolutionized the electronics industry in computers, information processing, and communications.

Now, however, semiconductors are being used more and more where considerable "brawn" is required. Devices such as high-power bipolar junction transistors and power field-effect transistors, as well as SCRs, TRIACs, GTOs, and other semiconductor switching devices that use a *p-n-p-n* regenerative effect to achieve bistable action, are expanding the power-handling horizons of semiconductors and finding increasing application in a wide range of products including regulated power supplies, lamp dimmers, motor drives, pulse modulators, and heat controls. HVDC and electric-vehicle propulsion are two additional areas of application which may have a very significant long range impact on the technology. The impact of solid-state devices capable of handling appreciable power levels has yet to be fully realized.

Since it first became available in late 1957, the SCR or silicon-controlled rectifier (also called the reverse blocking triode thyristor) has become the most popular member of the thyristor family. At present, SCRs are available from a large number of manufacturers in this country and abroad. SCR ratings range from less than one ampere to over three thousand amperes with voltage ratings in excess of three thousand volts.

For relatively low-power ac voltage control, the TRIAC is widely used. The majority of these bidirectional triode thyristors are rated at or below 40 A and at voltages up to 600 V. Low-cost plastic packages are being used for both the SCR and the TRIAC.

The gate turn-off thyristor (GTO) has been of considerable interest since the first SCR was produced. Although low-current GTO devices were fabricated more than twenty-five years ago, high-power devices were not available until much more recently. Japanese semiconductor manufacturers now lead in this technology, and they manufacture production devices with ratings up to 2500 V, 2000 A.

Other special thyristor devices are also being produced. These include light-activated silicon-controlled rectifiers, asymmetrical SCRs, reverse-conducting triode thyristors, bidirectional diode thyristors, and reverse-blocking diode thyristors.

Continuing progress has also been made in high-power bipolar-transistor technology. At present, a wide range of silicon devices are available. Essentially all current developments involve silicon power components. Japanese manufacturers produce devices with ratings up to 1000 V and several hundred amperes. The Westinghouse D60T, D62T, and D7S series presently are the highest-power devices available in the United States.

Rapid developments are occurring in power field-effect transistors. They show considerable promise for high-frequency application because of their very fast switching speeds, and at lower frequency as well, because of the relative simplicity of the control-circuit requirements. However, at present the highest rating devices are not in the same power range as thyristors and bipolar junction transistors. For this reason, this book does not include circuit applications of power FETs.

Another power FET-type device which is presently being developed in Japan is the static induction transistor. The insulated-gate transistor (IGT) now available from General Electric has the input characteristics of the FET but the low saturation voltage of the bipolar transistor. Thus it is a promising new component.

Some related devices of interest to the solid-state power circuit engineer include unijunction transistors, silicon unilateral and bilateral switches, low-cost phototransistors, light-emitting diodes, opto-isolators, zener or avalanche diodes, fast-recovery diodes with ratings from 10 A to over 1000 A, and semi-conductor chips.

Although the integrated circuit is presently of most use in gating and control of power semiconductor circuits, some hybrid power modules are being produced. These include IC and discrete components to perform a complete circuit function.

Another extremely important current development is the application of microprocessors for power electronic control circuits to replace discrete and integrated electronic devices. With a modern microprocessor, it is feasible to produce very sophisticated controls at reasonable cost.

All of these developments have led to the use of power semiconductors in a rapidly expanding range of applications.

Power electronics is a fast-moving technology, which makes it difficult to produce a text that is not obsolete at the time of printing. The strategy adopted herein to deal with this difficulty is to concentrate on the more well-known and widely applied circuits and devices. A truly up-to-date text should include a chapter on GTO device characteristics but, regrettably, this chapter could not be completed in time to meet the text's press deadline. Also, the gating and control circuits are extremely important in the design of reliable systems, but this could be the subject of another text on signal-level electronics for power applications.

This book is written as an introductory text for electrical engineering students and practicing engineers interested in power electronics. It is designed to provide the student with knowledge of the important circuit techniques possible using bipolar-junction power transistor and thyristor devices. For the past

10 years, the author has taught a two-semester course sequence on this subject at the University of Missouri–Columbia. The notes for these courses are the basis for this text. The text is considered appropriate for a four-credit-hour one-semester course, including a laboratory, for senior electrical engineering students, followed by a three-credit-hour senior or first-year graduate course.

I wish to express my appreciation to a number of persons who helped make this text possible — to Bernie Bedford and my other associates at General Electric early in my career, to my colleagues in the UMC Department of Electrical and Computer Engineering for additional work they may have done because I was devoting so much effort to the preparation of this text, to the students in my classes for helping to correct the draft manuscript, to Suang Khuwatsamrit and Romney Katti for reviewing portions of the final draft, to my colleague Atsuo Kawamura for his considerable assistance, to Carla George for an outstanding job of typing the final manuscript, and to my wife for her patience, understanding, and support throughout my work on this project.

RICHARD G. HOFT

SYMBOLS

The symbols are defined in the text as they are used. A number of the more important ones are listed below as an aid to the reader.

E	Constant dc voltage, e.g., battery or machine counter electromotive force (EMF)
i or $i(t)$	Instantaneous current
I	Average value of a periodic current waveform, or half-cycle average value of an ac current
I_e	Effective or RMS value of periodic current
I_m	Maximum or crest value of a sinusoidal ac current
$I_{m,n}$	Maximum or crest value of nth harmonic component of a periodic current waveform
ϕ	Power factor angle, i.e., the angle whose cosine is the power factor
p or $p(t)$	Instantaneous power
P	Power, or average power, or active power (watts)
PF	Power factor, i.e., the ratio of the total watts to the total RMS volts multiplied by the RMS amperes. (Note: The term displacement power factor is the corresponding ratio of the fundamental power to the fundamental RMS volts multiplied by the fundamental RMS amperes.)
S	Phasor power, i.e., the phasor voltage times the conjugate of the phasor current. (Sinusoidal quantities are assumed and the amplitude of the phasor is the RMS value of the corresponding sinusoidal quantity.)
VA	Apparent power, i.e., the RMS volts multiplied by the RMS amperes
VAR	Volt-amps-reactive, i.e., the RMS value of a sinusoidal voltage times the RMS value of a sinusoidal current times the sine of the angle by which the current lags the voltage. (Note: The VARs are positive when the current lags the voltage.)
v or $v(t)$	Instantaneous voltage
$v_{a\text{-}o}$	Instantaneous voltage of point a with respect to o, where positive voltage occurs when a is positive with respect to o
V	Average value of a periodic voltage waveform, or half-cycle average value of an ac voltage
V_e	Effective or RMS value of periodic voltage
V_m	Maximum or crest value of a sinusoidal ac voltage
$V_{m,n}$	Maximum or crest value of nth harmonic component of a periodic voltage waveform

CONTENTS

1
INTRODUCTION

1.1. HISTORY

Power electronics originated at the beginning of this century. In the chronological list of references "Bibliography on Electronic Power Converters," published by the American Institute of Electrical Engineers in February 1950, the first citation is for the year 1903. Many technical articles and several books on the subject were published during the period from 1930–1947. These dealt primarily with the application of grid-controlled gas-filled tubes. Because of the limitations of the mercury-arc rectifier and gas-filled thyratrons, only a relatively limited number of equipments were manufactured.

The invention of the bipolar junction transistor in 1948 was the beginning of semiconductor electronics. This device and semiconductor diodes spawned a revolution in the electronics industry. Drastic reductions in size, cost, and power consumption were achieved simultaneously with greatly increased equipment complexity and capability.

Semiconductor power diodes became available shortly after 1950. However, it was not until late 1957, when the most popular member of the thyristor family—the SCR—was announced by General Electric, that semiconductor power electronics really began. Starting from a single 16-A device, the thyristor family has grown tremendously. Hundreds of thyristor devices are available from numerous manufacturers throughout the world. In addition, there have been continuing improvements in ratings and characteristics of power transistors so that a wide variety of these components are also now available.

1.2. DEVICES AND TRENDS

Table 1-1 is a list of a number of thyristor devices presently available.

A wide range of power-switching transistors is also available, with current ratings up to several hundred amperes and voltage ratings well over 1000 V. However, the combination of the highest current and voltage rating is not available in a single device. One of the highest power devices presently available is the Westinghouse D60T/D62T. These are *npn* power-switching transistors with a maximum collector current of 200 A, a maximum $V_{CEO(SUS)}$ of 500 V and with a current gain of 10 at 50-A collector current.

There are many additional special thyristor and transistor devices. These include multiple bipolar transistors fabricated in a single package, chip-type (unpackaged) thyristors, power FETs, and Schottky diodes. In addition, the continued development of IC components and microprocessors is having a most important impact on control circuits for power electronic systems.

Table 1-1. Thyristor Devices

Asymmetrical silicon bilateral switch (ASBS)
Asymmetrical silicon-controlled rectifier (ASCR)
Bidirectional diode thyristor (DIAC)
Bidirectional triode thyristor (TRIAC)
Gate-assisted turn-off thyristor (GAT or GATT)
Gate turn-off thyristor (GTO)
Light-activated programmable unijunction transistor (LAPUT)
Light-activated reverse-blocking diode thyristor (LAS — light-activated switch)
Light-activated reverse-blocking tetrode thyristor (LASCS)
Light-activated reverse-blocking triode thyristor (LASCR)
Programmable unijunction transistor (PUT)
Reverse-blocking diode thyristor (RBDT)
Reverse-blocking tetrode thyristor (SCS — silicon-controlled switch)
Reverse-blocking triode thyristor (SCR)
Reverse-conducting diode thyristor (RCDT)
Reverse-conducting triode thyristor (RCTT)
Silicon bilateral switch (SBS)
Silicon unilateral switch (SUS)

Device trends of most importance at this time include the following. There is continuing growth in the number of switching power transistors with a combination of high voltage and current ratings. This is causing an increase in the rating level at which power transistor equipments are preferable to thyristor hardware. For the past several years, there has been considerable activity on the development of high-power gate turn-off devices with improved characteristics. These components could have great application potential. Presently, at least four Japanese companies manufacture GTOs with ratings up to 2500 V, 1000 A — Hitachi, Mitsubishi, Toshiba, and IR-Japan. A third area of interest is the activity on improved cooling techniques for power semiconductors. New types of heat sinks integral with the device have significant potential for reducing size and cost per unit of equipment rating. More power FETs are certain to become available, a greater number of power integrated devices is anticipated, and there is considerable opportunity for device improvement with clever new packaging techniques.

The impact of microprocessors on power electronics cannot be overemphasized. With the availability of low-cost, very sophisticated microprocessors, hardware which previously was not considered because of its control complexity now is feasible. This is greatly expanding the horizons for sophisticated power electronics systems.

Table 1-2 is a list of domestic manufacturers of power semiconductor components.

Table 1-2. Power Semiconductor Device Manufacturers (U.S.A.)

Atlantic Semiconductors/Diodes
Edal Industries
EE Tech
Electronic Devices
Fairchild Semiconductor
General Electric
General Instrument
General Semiconductor Industries
Hewlett-Packard
Hughes
International Rectifier
Intersil
Lansdale
Motorola Semiconductor Products
Power Physics
Power Semiconductors, Inc.
Power Tech
RCA
Sarkes Tarzian
Semtech Corporation
Solitron Devices
Solid State Devices
Teccor Electronics
Texas Instruments
TRW
Tungsol
Unitrode Corporation
Varo, Inc.
Westcode
Westinghouse Electric

1.3. APPLICATIONS

Table 1-3 is a list of applications by product type. This is not intended to be all inclusive but rather to illustrate the spectrum of uses of power semiconductors in commercially available products.

Table 1-4 is a listing of typical products by function.

1.4. UNIQUE ASPECTS OF POWER ELECTRONICS

1.4.1. Switching Operation

When semiconductors are used in high-power applications, they are generally operated as switches. There are only a very few exceptions to this, such

Table 1-3. Power Semiconductor Applications

AEROSPACE AND DEFENSE
  Aircraft power supplies
  Laser power supplies
  Radar/sonar power supplies
  Solid-state relays, contactors,
    and circuit breakers
  Sonobuoy flashers
  Space power supplies
  VLF transmitters

CONSUMER
  Audio amplifiers
  Electric door openers
  Heat controls
    Electric blanket
    Electric dryer
    Food-warmer tray
    Furnace
    Oven
    Range surface unit
  High-frequency lighting
  Light dimmers
  Light flashers
  Motor controls
    Air conditioning
    Blender
    Electric fan
    Food mixer
    Garage-door opener
    Hand power tool
    Model train
    Movie projector
    Sewing machine
    Slot car
  RF amplifiers
  Security systems
  TV deflection

INDUSTRIAL
  Mercury-arc lamp ballasts
  Motor drives
    Cement kiln
    Conveyor
    Crane and hoist
    Machine tool
    Mining
    Oil-well drilling
    Paper mill
    Printing press
    Pump and compressor
    Steel mill
    Synthetic fibre

INDUSTRIAL (continued)
  Power supplies
    Aluminum reduction
    Battery charger
    Computer
    Electrochemical
    Electroplating
    Electrostatic precipitator
    Induction heating
    Laboratory
    Mining
    Particle accelerator
    Welding
  Static relays and circuit breakers
  Ultrasonic generators

POWER SYSTEMS
  Gas turbine starting
  Generator exciters
  HVDC
  Nuclear-reactor control-rod drives
  Solar power supplies
  Synchronous machine starting
  Uninterruptible power supplies
  VAR compensation
  Wind generator converters

TRANSPORTATION
  Electronic ignition
  Linear induction motor control
  Motor drives
    Electric vehicle
    Elevator
    Fork-lift truck
    Locomotive
    Mass transit
    People movers
  Traffic-signal control
  Voltage regulator

Table 1-4. Power Semiconductor Application Functions

STATIC SWITCHING
  Solid-state relays, contactors, and circuit breakers
  Logic systems
  Circuit protectors — crowbars, limit activated interrupters

ac PHASE CONTROL
  Light dimmers
  Motor speed controls
  Voltage regulators
  VAR regulators

PHASE-CONTROLLED RECTIFIER/INVERTER
  dc motor drives
  Regulated dc power supplies
  HVDC
  Wind generator converters

CYCLOCONVERTER
  Aircraft VSCF systems
  Variable-frequency ac motor drives

FREQUENCY MULTIPLIER
  Induction-heating supplies
  High-frequency lighting

TRANSISTOR LINEAR AMPLIFIER
  Audio amplifiers
  RF amplifiers

TRANSISTOR SWITCHING REGULATOR
  dc-dc buck, boost, and buck-boost converters
  High-performance regulated power supplies

THYRISTOR CHOPPER
  Electric transportation propulsion control
  Generator exciters
  High-performance, high-power regulated supplies

dc-ac INVERTER
  Aircraft and space power supplies
  Uninterruptible power supplies
  Variable-frequency ac motor drives

as linear amplifiers and series or shunt transistor-regulated power supplies. When the semiconductor is used as a switch, it is possible to control large amounts of power to a load with a relatively low power dissipation in the switching device. For example, in an ideal switch with zero voltage drop when "on," zero leakage current when "off," and zero switching time, there is no power dissipation in the switching device regardless of the on-state current or off-state voltage, and there is no switching loss. In a practical case, the on-state voltage drop across the switching device limits the on-state current because of device dissipation limits. All practical power semiconductor devices also have a limited off-state voltage rating. When high-frequency switching is employed, there also may be significant switching loss due to the large instantaneous device power dissipation because of the finite switching times of practical power semiconductors.

Figure 1-1 illustrates the switching action of a simple switching transistor circuit. When the transistor is off or open, the collector-emitter voltage is assumed to be E volts. Actually, the off-state transistor voltage would be slightly less than the dc supply voltage because of the load voltage drop due to the leakage current. The collector current flowing when the transistor is on is assumed to be I amperes. In addition, the instantaneous collector-emitter voltage $v(t)$ and collector current $i(t)$ are both assumed to change in a linear fashion during the switching interval. Although this is not exactly the case in a practical situation,

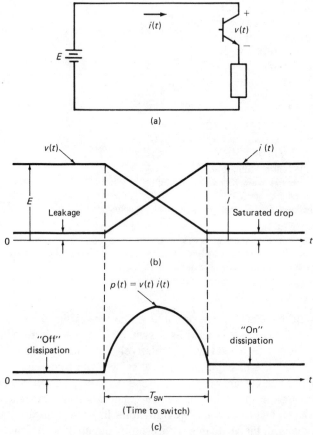

Fig. 1-1. Transistor switching.

the linear variations are a reasonably good approximation, and they most clearly illustrate the important switching characteristics.

The instantaneous power dissipated during the switching interval can be expressed as

$$P_T = v(t)i(t) = \left[\frac{E(T_{SW} - t)}{T_{SW}}\right]\left[I\frac{t}{T_{SW}}\right] = \frac{EI}{(T_{SW})^2}(T_{SW} - t)t \qquad (1\text{-}1)$$

In the expressions for $v(t)$ and $i(t)$, the beginning of the switching interval is assumed to be $t = 0$. Also, the saturated voltage drop and the collector leakage current are both assumed to be negligible.

The average power dissipated during a switching interval is important since it determines the maximum number of switchings possible in a given time interval. The average dissipation during the interval T_{SW} is given by

$$P_{TSW} = \frac{1}{T_{SW}} \int_0^{T_{SW}} v(t)i(t)\, dt \qquad (1\text{-}2)$$

Combining with Eq. (1-1),

$$P_{TSW} = \frac{EI}{(T_{SW})^3} \int_0^{T_{SW}} (T_{SW} - t)t\, dt = \frac{EI}{(T_{SW})^3}\left[\frac{(T_{SW})^3}{2} - \frac{(T_{SW})^3}{3}\right] = \frac{EI}{6} \qquad (1\text{-}3)$$

The total average dissipation in a switching element is obtained by adding the on-state, off-state, and switching losses. For example, with a switching period of T, assuming linear switching with a switching time of T_{SW} and an on- and off-time of T_{ON} and T_{OFF}, respectively

Total device average dissipation $\triangleq P_T$

$$P_T \approx \frac{2(EI/6)T_{SW} + (V_{CE(SAT)}I)T_{ON} + (EI_{leakage})T_{OFF}}{T} \qquad (1\text{-}4)$$

where $V_{CE(SAT)}$ and $I_{leakage}$ are assumed small enough such that the switching loss is approximately given by Eq. (1-3).

1.4.2. Commutation

The *IEEE Standard Dictionary of Electrical and Electronic Terms* [1] contains the following definition of commutation:

The transfer of unidirectional current between rectifier circuit elements or thyristor convertor circuit elements that conduct in succession.

As indicated, the term commutation was originated to define a process occurring in rectifier and thyristor converters. In general, complete commutation may involve a number of events. The most important of these are (1) the reduction of forward current to zero in one power semiconductor switching element, (2) the delay of reapplication of forward voltage to this element until it has regained its forward-blocking capability, (3) the provision of an alternate path for the load-current flow until the next element to conduct is turned on, and (4) the build-up of forward current in the next element which is to conduct. These events may occur concurrently or in sequence. Reliable operation of most power electronic equipment is critically dependent upon how well the commutation process is handled.

When power transistors or GTO devices are used as the switching elements, the commutation process is simplified. The current flow stops when the transistor base drive is removed or driven negative. Also, the transistor will almost immediately revert to the blocking or off-state when this occurs. A GTO regains its blocking capability in the order of ten microseconds after the required negative gate current is applied. However, it is still very important to provide an alternate path for load-current flow during transistor or GTO switching when the load is inductive.

1.4.3. Most Important Device Characteristics

When semiconductors are used for power applications, the device characteristics of greatest importance are the on-state voltage drop, the on-state current-handling ability, the off-state blocking-voltage capability, the switching speed, the rate of recovery of blocking capability, the power-dissipation limits, and the power gain. There are additional more detailed characteristics of particular power semiconductor devices which are important, but those of most general importance are listed in the preceding sentence. The number of devices required for a given application is determined by the device current, voltage, and power-dissipation ratings. With a higher on-state voltage drop, the device dissipation increases, which is the basic limitation on current-handling capability. If the device switching speeds are low, considerable average dissipation will result when operating at high switching frequency. The switching frequency is also limited by the rate of recovery of blocking capability. Finally, the power gain of the switching device determines the drive requirements and thus the size, cost, and complexity of the power electronic control.

1.5. IMPORTANT PRINCIPLES

In Appendix I of the excellent text by the late William E. Newell [2], he lists "ten cornerstones of power electronics." The first two are Kirchoff's voltage and current laws. The next three are the v-i relations for the resistor (Ohm's law), inductor, and capacitor. These five "cornerstones" should be basic tools of every electrical engineer. The remaining five "cornerstones" of Newell are briefly discussed here because they are not quite as well understood by every electrical engineer.

1.5.1. Average and RMS

The average and root-mean-square (RMS) values are very important characteristics of electrical waveforms, as well as of the many other physical quantities. They are determined from the following expressions:

$$[f(t)]_{\text{AVE}} = \frac{1}{T} \int_{t_0}^{T+t_0} f(t)\, dt \tag{1-5}$$

$$[f(t)]_{\text{RMS}} = \sqrt{\frac{1}{T} \int_{t_0}^{T+t_0} [f(t)]^2\, dt} \tag{1-6}$$

These expressions are valid for any arbitrary function $f(t)$ with the one constraint that the function must be periodic. The period is denoted by the symbol T.

1.5.2. Power

The instantaneous power is the product of the instantaneous values of the voltage and current.

$$p(t) = v(t)i(t) \tag{1-7}$$

A general expression for the calculation of the average power (also called active power, or just power) is

$$P = \frac{1}{T} \int_{t_0}^{T+t_0} vi\, dt \tag{1-8}$$

In this book, the quantity defined by Eq. (1-8) is called simply "power." It is the average value of the instantaneous power. The instantaneous voltage and current can be arbitrary waveforms, but both must be periodic with the same period, T.

Most engineers are very familiar with the expressions for power in dc and in sinusoidal circuits. For a battery with a constant voltage of E volts and a steady current through the battery of I amperes, the power is

$$P = \frac{1}{T} \int_{t_0}^{T+t_0} EI\, dt = EI \tag{1-9}$$

Since both the voltage and the current are constants, E and I can be moved outside the integral. The integral is then trivial to perform, yielding the result in Eq. (1-9). In this book, the power delivered from a battery is considered positive. If the battery current is reversed, implying that the battery is being charged, the power now supplied to the battery is given a negative sign.

The power dissipated in a resistor is also simple to determine using the fundamental expression in Eq. (1-8), the v-i relation for the resistor, and the expression for RMS, Eq. (1-6). Consider an arbitrary current $i(t)$ flowing through a resistor R. Then

$$P = \frac{1}{T} \int_{t_0}^{T+t_0} [i(t)R][i(t)] \, dt = R \frac{1}{T} \int_{t_0}^{T+t_0} [i(t)]^2 \, dt = RI_{\text{RMS}}2 = RI_e^2$$

$$(1\text{-}10)$$

(Note: The subscript e is used to denote the RMS or effective value.)

In many power electronic circuits, both $v(t)$ and $i(t)$ may be distorted and time-varying waveforms. In these situations, it is necessary to use an "electronic wattmeter" to measure power. Such an instrument contains a multiplier to determine the product $[v(t)][i(t)]$. Then, it also contains a cyclic integrator to compute the average value in Eq. (1-8). The frequency response of the multiplier must be high enough such that all significant frequency components in $[v(t)][i(t)]$ are included without appreciable attenuation.

If it is desired to determine the copper loss in winding resistances, it is important to measure the RMS current or voltage properly in cases where the waveforms are general time-varying functions. In most power electronic circuits, abrupt switchings occur, and these result in quite high-frequency components in the waveforms. A thermal meter of some type is frequently used for RMS measurements. There are also electronic instruments now available which perform the computations necessary to determine the RMS value given by Eq. (1-6). Again the frequency response of such instruments is critical, so that all significant frequency components in the current or voltage waveform are included in the determination of the RMS quantity.

1.5.3. Steady-State Conditions for Inductor and Capacitor

There are two very important steady-state principles which are most useful in the analysis of power electronic circuits. The first of these is that *the average voltage across an inductor over a full cycle in steady state is zero*. This is quite easy to show using the v-i relation for an inductor, the definition of steady state, and the basic expression in Eq. (1-5) for the average value. Consider an inductor L, with a voltage across it of $v_L(t)$ and with a current flowing through it of $i_L(t)$:

$$i_L(t) = \frac{1}{L} \int v_L(t) \, dt \tag{1-11}$$

or

$$i_L(t_0 + T) - i_L(t_0) = \frac{1}{L} \int_{t_0}^{t_0+T} v_L(t) \, dt \tag{1-12}$$

By definition, in steady state the current at the end of a period, $i_L(t_0 + T)$, must be equal to the current at the beginning of the period, $i_L(t_0)$. Thus,

$$\frac{1}{L} \int_{t_0}^{t_0+T} v_L(t) \, dt = 0 \tag{1-13}$$

If Eq. (1-13) is multiplied by the constant, L/T, then the left side of this equation is the average voltage across the inductor. Since the right side remains equal to zero, the principle *the average voltage across an inductor over a full cycle in steady state is zero* is shown.

The second important steady-state principle is that *the average current through a capacitor over a full cycle in steady state is zero*. This can be shown in a similar manner, considering a capacitor C with a voltage across it of $v_c(t)$ and a current through it of $i_c(t)$.

$$v_c(t) = \frac{1}{C} \int i_c(t) \, dt \tag{1-14}$$

or

$$v_c(t_0 + T) - v_c(t_0) = \frac{1}{C} \int_{t_0}^{t_0+T} i_c(t) \, dt \tag{1-15}$$

In steady state, the voltage across the capacitor at the end of a period, $v_c(t_0 + T)$, must equal the voltage at the beginning of the period, $v_c(t_0)$. Therefore

$$\frac{1}{C} \int_{t_0}^{t_0+T} i_c(t) \, dt = 0 \tag{1-16}$$

If this equation is multiplied by the constant C/T, the left side becomes the average current. Since the right side remains zero, the principle *the average current through a capacitor over a full cycle in steady state is zero* is shown.

EXAMPLE 1-1

Consider the switching circuit in Fig. 1-2 with $E = 200$ V and $R = 10$ Ω.*
Assume that the periodically operated switch is closed for 10 μs and opened for
40 μs, and that the inductor is large enough such that its current is always
greater than zero during both the on- and off-intervals of the switch. Use the
steady-state principles for the inductor and capacitor to calculate the average
current through the resistor R and through the inductor L in steady state.

Solution:

When the switch is closed, the diode is reverse-biased and thus the voltage v_o
is equal to E. The inductor current i_L is assumed to be greater than zero for all
time during each switching period in steady state. When the switch is opened,
the inductor current cannot change abruptly, so that it immediately flows
through the diode. During the entire off-interval of the switch, the current i_L
flows through the diode, which is referred to as a coasting or freewheeling
diode. The diode current decays during the freewheeling interval with a time
function determined by the particular E, R, L, C, switching period T, and the
switching duty cycle d. The duty cycle of the switch is defined as

$$d \triangleq \frac{\text{on-time of switch}}{\text{switching period}} = \frac{t_{ON}}{T} \qquad (1\text{-}17)$$

The average value of the voltage v_o is

$$V_o = \frac{1}{T} \int_{t_0}^{t_0+T} v_o(t)\, dt = \frac{1}{T} \int_{t_0}^{t_0+t_{ON}} (E)\, dt = \frac{t_{ON}}{T} E = \left(\frac{10}{50}\right)(200) = 40 \text{ V}$$

$$(1\text{-}18)$$

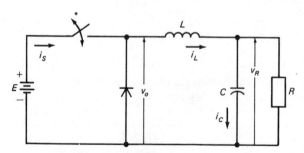

*This symbol is used for a repetitively
operated switch throughout this text

Fig. 1-2. Circuit for Example 1-1.

*Throughout this text, ideal components will be assumed unless specifically stated otherwise.

In steady state, the average voltage across the inductor must be zero. Thus, the average value of v_o must all appear across the resistor. Therefore, the average current through R is

$$I_R = \frac{V_R}{R} = \frac{V_o}{R} = \frac{40}{10} = 4 \text{ A} \tag{1-19}$$

Since no average current can flow through the capacitor in steady state, the average value of the inductor current is equal to the average current through R.

$$I_L = I_R = 4 \text{ A} \tag{1-20}$$

The inductance and capacitance values would have to be known in order to determine the specific time functions for the circuit voltages and currents. It should be noted that i_L and v_R are not composed of simple exponential functions of time, since they are given by the solutions of the second-order differential equations representing the circuit during the on- and off-intervals.

EXAMPLE 1-2

Consider the circuit of Fig. 1-2 again with $E = 200$ V, $R = 10 \ \Omega$, $t_{ON} = 10 \ \mu s$, and $d = 0.2$. However, now assume that the capacitor is so large that there is negligible ripple in the load voltage v_R and assume an inductance $L = 400 \ \mu H$. Determine the steady-state waveforms for i_S, i_L, and i_C.

Solution:

Since the supply voltage E and the duty cycle d are the same as in the previous example, $V_o = 40$ V and $I_R = 4$ A. With the assumption of a large enough capacitor such that there is negligible ripple in the load voltage v_R, this voltage is essentially a constant, as shown in Fig. 1-3(a). Now the slope of the current through the inductor is determined from $v = L(di/dt)$ as follows:

$$\frac{di_L}{dt} = \frac{v_o - v_R}{L} \tag{1-21}$$

$$\left. \frac{di_L}{dt} \right|_{\text{on-interval}} = \frac{E - v_R}{L} = \frac{200 - 40}{400 \ \mu H} = 0.4 \text{ A}/\mu s \tag{1-22}$$

$$\left. \frac{di}{dt} \right|_{\text{off-interval}} = \frac{0 - 40}{400 \ \mu H} = -0.1 \text{ A}/\mu s \tag{1-23}$$

Since the two triangular portions of the current i_L have the same height, the average value of i_L is the same during the on-interval and during the off-

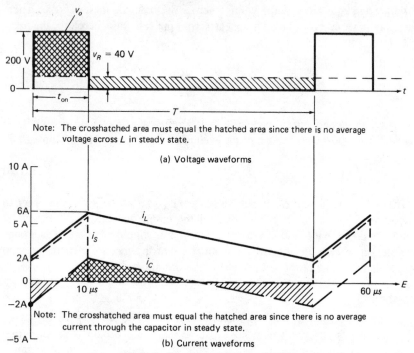

Note: The crosshatched area must equal the hatched area since there is no average voltage across L in steady state.

(a) Voltage waveforms

Note: The crosshatched area must equal the hatched area since there is no average current through the capacitor in steady state.

(b) Current waveforms

Fig. 1-3. Waveforms for Example 1-2.

interval. The average value of the inductor current must also equal the average of the current through the resistor R, as there can be no average current through the capacitor in steady state. Figure 1-3(b) shows the steady-state current waveforms.

A useful check on the results for this example is to calculate the power delivered from the dc supply, which must equal the power delivered to the load resistor.

$$P_S = EI_S = 200 \text{ V} \times 0.8 \text{ A} = 160 \text{ W} \qquad (1\text{-}21)$$

$$P_R = \frac{V_{R,e}^2}{R} = \frac{(40)^2}{10} = 160 \text{ W} \qquad (1\text{-}22)$$

The average value of the dc supply current is obtained easily from the waveform for i_S in Fig. 1-3(b). It is very important to note that in this case the voltage across R is essentially a constant. Thus

$$V_R = V_{R,e} = v_R = 400 \text{ V} \qquad (1\text{-}23)$$

In a more general case, with ripple in the voltage v_R, it would be necessary to calculate the RMS value of v_R to determine the power delivered to the resistor.

1.6. HARMONIC ANALYSIS

In most power electronic circuits the voltage and current waveforms are distorted and nonsinusoidal. Thus harmonic analysis is a very useful analytical tool for the power electronic engineer. Many textbooks have excellent treatments of this subject, e.g., [3] and [4]. A brief review of the essential elements of harmonic analysis is presented in this section.

All periodic waveforms arising in physical systems can be represented by a Fourier series of the form

$$f(\omega_1 t) = A_o + A_1 \cos \omega_1 t + A_2 \cos 2\omega_1 t + A_3 \cos 3\omega_1 t + \ldots$$

$$+ B_1 \sin \omega_1 t + B_2 \sin 2\omega_1 t + B_3 \sin 3\omega_1 t + \ldots$$

$$= A_o + \sum_{n=1}^{\infty} (A_n \cos n\omega_1 t + B_n \sin n\omega_1 t) \tag{1-24}$$

where $\omega_1 = 2\pi/T$ is the fundamental frequency with period T, n takes on all positive integer values, and the A_n's and B_n's are the amplitudes of the cosine and sine terms in the Fourier series. The Fourier representation is based on the very useful property of sinusoidal functions — that every arbitrary $f(\omega_1 t)$ for a physical system can be represented by the summation of a set of sine and cosine waves. In general, an infinite number of sine and cosine waves are required, each having frequencies which are an integer multiple of the fundamental frequency and each having unique amplitudes, A_n and B_n. It is worthy of note that $f(\omega_1 t)$ is restricted to functions arising in physical systems. The mathematician is always capable of defining a function which does not permit the application of even quite generally accepted mathematical operations. Fortunately, functions arising in physical systems are single-valued and have only a finite number of discontinuities in any finite interval, which together are sufficient conditions for convergence of the Fourier series.

The equations which are used to determine the coefficients of the terms in the Fourier series are as follows:

$$A_o = \frac{1}{T} \int_0^T f(t) \, dt \tag{1-25}$$

$$A_n = \frac{2}{T} \int_0^T f(t) \cos n\omega_1 t \, dt \tag{1-26}$$

$$B_n = \frac{2}{T} \int_0^T f(t) \sin n\omega_1 t \, dt \tag{1-27}$$

where $n = 1, 2, 3, \ldots$ (all positive integers).

Equation (1-25) is the average value or dc component of the waveform $f(t)$. These equations can also be written in terms where the variable of integration is the angle $\omega_1 t$, as follows:

$$A_o = \frac{1}{2\pi} \int_0^{2\pi} f(\omega_1 t) \, d(\omega_1 t) \tag{1-28}$$

$$A_n = \frac{1}{\pi} \int_0^{2\pi} f(\omega_1 t) \cos n\omega_1 t \, d(\omega_1 t) \tag{1-29}$$

$$B_n = \frac{1}{\pi} \int_0^{2\pi} f(\omega_1 t) \sin n\omega_1 t \, d(\omega_1 t) \tag{1-30}$$

Generally, these latter equations are more convenient to apply.

The coefficient equations are easily derived from Eq. (1-24) by multiplying both sides of this relation by an appropriate function and then integrating over the full period. For example, if both sides of Eq. (1-24) are multiplied by one and then integrated over the period, the A_0 expression is obtained:

$$\int_0^{2\pi} f(\omega_1 t) \, d(\omega_1 t) = \int_0^{2\pi} \left[A_o + \sum_{n=1}^{\infty} (A_n \cos n\omega_1 t + B_n \sin n\omega_1 t) \right] d(\omega_1 t)$$

$$= \int_0^{2\pi} A_o \, d(\omega_1 t) = (2\pi)(A_0) \tag{1-31}$$

which yields Eq. (1-28). In performing the integration, an important property of the sinusoidal wave has been used; the integral over the period is zero for every sine or cosine function whose frequency is a positive integer multiple of the fundamental frequency. In fact, all of the following definite integrals for sinusoidal functions are useful in the derivation of Eqs. (1-28), (1-29), and (1-30):

$$\int_0^{2\pi} \cos n\omega_1 t \, d(\omega_1 t) = \int_0^{2\pi} \sin n\omega_1 t \, d(\omega_1 t) = \int_0^{2\pi} \sin n\omega_1 t \cos n\omega_1 t \, d(\omega_1 t) = 0 \tag{1-32}$$

$$\int_0^{2\pi} \cos^2 n\omega_1 t \, d(\omega_1 t) = \int_0^{2\pi} \sin^2 n\omega_1 t \, d(\omega_1 t) = \pi \tag{1-33}$$

where n is any positive integer. If m and n are both positive integers but with $m \neq n$, then

$$\int_0^{2\pi} \cos m\omega_1 t \, \sin n\omega_1 t \, d(\omega_1 t) = \int_0^{2\pi} \cos m\omega_1 t \, \cos n\omega_1 t \, d(\omega_1 t)$$

$$= \int_0^{2\pi} \sin m\omega_1 t \, \sin n\omega_1 t \, d(\omega_1 t) = 0 \qquad (1\text{-}34)$$

It is quite useful in the determination of the coefficient for the Fourier series to use symmetry properties of the distorted waveforms. The following three types of symmetry occur often in practical waveforms:

1. Odd-function symmetry; $f(\omega_1 t) = -f(-\omega_1 t)$. Waveforms with this symmetry contain only sine terms.
2. Even-function symmetry; $f(\omega_1 t) = f(-\omega_1 t)$. Waveforms with this symmetry contain only cosine terms.
3. Half-wave symmetry; $f(\omega_1 t) = -f(\omega_1 t + \pi)$. Waveforms with this symmetry contain only odd harmonics.

EXAMPLE 1-3

Derive (1-29).

Solution:

Both sides of Eq. (1-24) are multiplied by $\cos m\omega_1 t$ and integrated over the period

$$\int_0^{2\pi} f(\omega_1 t) \cos m\omega_1 t \, d(\omega_1 t) = \int_0^{2\pi} \left[A_0 + \sum_{n=1}^{\infty} (A_n \cos n\omega_1 t + B_n \sin n\omega_1 t) \right]$$

$$\cos m\omega_1 t \, d(\omega_1 t) = \int_0^{2\pi} A_o \cos m\omega_1 t \, d(\omega_1 t)$$

$$+ \int_0^{2\pi} \sum_{n=1}^{\infty} (A_n \cos n\omega_1 t) \cos m\omega_1 t \, d(\omega_1 t)$$

$$+ \int_0^{2\pi} \sum_{n=1}^{\infty} (B_n \sin n\omega_1 t) \cos m\omega_1 t \, d(\omega_1 t)$$

$$= \int_0^{2\pi} A_m \cos^2 m\omega_1 t \, d(\omega_1 t) = A_m \pi \qquad (1\text{-}35)$$

where the properties of Eqs. (1-32) through (1-34) have been used to simplify the right side. If both sides of Eq. (1-35) are divided by π, and then the sides are reversed,

$$A_m = \frac{1}{\pi} \int_0^{2\pi} f(\omega_1 t) \cos m\omega_1 t \, d(\omega_1 t) \qquad (1-36)$$

Then changing the integer symbol yields Eq. (1-29).

$$A_n = \frac{1}{\pi} \int_0^{2\pi} f(\omega_1 t) \cos n\omega_1 t \, d(\omega_1 t)$$

EXAMPLE 1-4

Determine the harmonics present in the square wave $v(\omega_1 t)$ shown in Fig. 1-4.

Solution:

With the $\omega_1 t = 0$ axis chosen as indicated in Fig. 1-4, this square wave has both even-function symmetry, $f(\omega_1 t) = f(-\omega_1 t)$, and half-wave symmetry, $f(\omega_1 t) = -f(\omega_1 t + \pi)$. Thus, the Fourier series will contain only odd harmonic, cosine terms.

From Eq. (1-29),

$$A_n = \frac{1}{\pi} \left[\int_0^{\pi/2} V \cos n\omega_1 t \, d(\omega_1 t) + \int_{\pi/2}^{3\pi/2} (-V) \cos n\omega_1 t \, d(\omega_1 t) \right.$$

$$\left. + \int_{3\pi/2}^{2\pi} V \cos n\omega_1 t \, d(\omega_1 t) \right]$$

$$= \frac{V}{\pi n} \left[\sin n\omega_1 t \Big|_0^{\pi/2} - \sin n\omega_1 t \Big|_{\pi/2}^{3\pi/2} + \sin n\omega_1 t \Big|_{3\pi/2}^{2\pi} \right]$$

$$= \frac{V}{\pi n} \left[\sin \frac{\pi n}{2} - \sin \frac{3\pi n}{2} + \sin \frac{\pi n}{2} - \sin \frac{3\pi n}{2} \right]$$

$$= \frac{2V}{\pi n} \left[\sin \frac{\pi n}{2} - \sin \frac{3\pi n}{2} \right] \qquad (1-37)$$

It is quite clear that Eq. (1-37) becomes zero for even integer values of n, since the argument of both terms becomes an even multiple of π. When n is an odd integer,

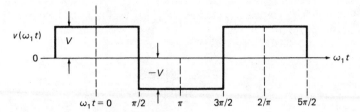

Fig. 1-4. Square wave with even-function symmetry and half-wave symmetry.

$$A_1 = \frac{4V}{\pi}$$

$$A_3 = -\frac{4V}{3\pi}$$

$$A_5 = \frac{4V}{5\pi} \qquad (1\text{-}38)$$

$$A_7 = -\frac{4V}{7\pi}$$

$$\vdots$$

$$A_n = (-1)^{(n+3)/2} \frac{4V}{n\pi}$$

Thus, the Fourier series representation for the square wave of Fig. 1-4 is

$$v(\omega_1 t) = \frac{4V}{\pi} \left(\cos \omega_1 t - \frac{1}{3} \cos 3\omega_1 t + \frac{1}{5} \cos 5\omega_1 t - \frac{1}{7} \cos 7\omega_1 t \right.$$

$$\left. + \frac{1}{9} \cos 9\omega_1 t - \dots \right) \qquad (1\text{-}39)$$

It should be noted that the integration to obtain Eq. (1-37) could have been simplified using the symmetry of the waveform. In this case, it is possible to integrate over only a quarter-cycle (0 to $\pi/2$) and multiply the result by four. However, in more complex situations, it is sometimes difficult to be certain that one can integrate over only a fraction of a cycle. Thus, in general it is safer to integrate over a complete cycle, and this does not add greatly to the complexity of the solution. The graphical technique of [5] is useful in the determination of the necessary integration interval.

REFERENCES

1. Jay, Frank, ed. *IEEE Standard Dictionary of Electrical and Electronic Terms*. New York: IEEE, Inc., 1977.
2. Newell, William E., and Motto, John W., Jr. *Introduction to Solid State Power Electronics*. Youngwood: Westinghouse Electric Corporation, 1977.
3. Scott, Ronald E. *Linear Circuits*. Reading: Addison-Wesley Publishing Company, 1960, pp. 679–700.
4. Skilling, Hugh H. *Electric Networks*. New York: John Wiley & Sons, 1974, pp. 309–337.
5. Bedford, B. D., and Hoft, R. G. *Principles of Inverter Circuits*. New York: John Wiley & Sons, 1964, pp. 280–286.

PROBLEMS

1. Generally, semiconductors are used in the switching mode for power electronic circuits, since this mode of operation makes it possible to control a large amount of load power with low power dissipation in the semiconductor switching device. This problem is intended to illustrate the rather large power dissipation in a transistor when it is used in the active mode as a linear amplifying element, instead of in the switching mode. Consider the circuit of Fig. 1P-1 with the supply voltage $E = 100$ V and a purely resistive load $R = 10$ Ω. Assume that a constant base current is supplied such that the collector current is I amperes. Derive expressions for the power delivered to the load and the power dissipated in the transistor as a function of E, R, and I. Calculate and plot the transistor dissipation and the power delivered to the load as a function of I, for values of $I = 0, 1, 2, \ldots, 10$ A. [Note: It is assumed that the transistor can be turned completely off such that there is negligible off-leakage current when $I = 0$. When $I = 10$ A, it is assumed that there is sufficient base current to drive the transistor hard into saturation such that the $V_{CE(SAT)}$ is essentially zero.]

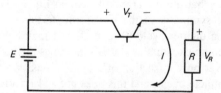

Fig. 1P-1. Series transistor regulator.

2. Repeat Problem 1. for the shunt transistor regulator shown in Fig. 1P-2, with the series resistor $R_S = 10\ \Omega$.

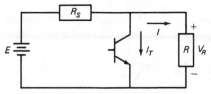

Fig. 1P-2. Shunt transistor regulator.

3. Consider the circuit of Fig. 1P-1 where the series transistor is operated as a switch at a fixed repetition frequency but with a varying duty cycle (on-time/switching period). Assume that the transistor on-state voltage drop is equal to 1 V and constant during the on-interval, but neglect the off-state leakage current and also neglect transistor switching losses. In this case, calculate and plot the power dissipated in the transistor and the power delivered to the load as a function of the duty cycle of the switching, with $E = 100$ V and $R = 10\ \Omega$.

4. Consider the switching transistor circuit of Fig. 1-1, where the transistor current and voltage are assumed to change linearly during switching. Assume that rated collector-emitter voltage is across the transistor when it is off, and that rated current flows through the transistor when it is on. These assumptions imply that both the off-state leakage current and the on-state voltage drop are negligible, and thus all transistor dissipation occurs during switching. The switching transistor ratings are as follows:

$$V_{CE} = 200\ V$$

$$I_C = 20\ A$$

Maximum average dissipation allowable $= 100$ W

(Neglect transistor dissipation due to base drive current)

$T_{SW} = 1\ \mu s$ (for both "on" and "off" switchings)

   a) Calculate and plot the instantaneous power dissipated in the transistor during a switching-on interval.
   b) Determine the maximum number of switchings per second allowable and the highest allowable frequency of operation for repetitive

switching. (Note: With repetitive switching of the circuit in Fig. 1-1, there is one on-switching followed by one off-switching during each cycle.)

5. Consider the switching transistor circuit of Fig. 1-1 again, but now assume that the transistor current and voltage change exponentially during switching. Assume negligible off-state leakage current and on-state voltage drop so that during a switching-on interval, the transistor voltage and current are given by the following expressions:

$$v(t) = E\varepsilon^{-t/\tau}; \qquad i(t) = I(1 - \varepsilon^{-t/\tau})$$

Calculate the transistor average-watts dissipation during a switching-on interval, considering the length of the interval to be 4τ seconds.

6. Consider the switching circuit of Fig. 1-1 during a switching-off interval in which

$$v(t) = E \quad \text{and} \quad i(t) = I\varepsilon^{-4\times10^6 t}$$

where E and I are constants. Assume that the switching-off interval is 1 μs. Calculate the average power dissipated in the transistor during this switching-off interval.

7. Calculate the average power dissipated in a transistor operated as a switch during the switching interval T_{SW}, where $v_{CE}(t)$ and $i_C(t)$ during the switching change as follows:

$$v_{CE}(t) = E$$

$$i_C(t) = I\left(1 - \frac{t}{T_{SW}}\right)$$

where E and I are constants.

8. For the simple switching circuit of Fig. 1P-3, assume that the ideal switch is repetitively operated with the on-time equal to one-fourth of the off-time.
   a) Calculate the average current through the load resistor.
   b) Calculate the power dissipated in the load resistor.

9. Consider the switching circuit of Fig. 1P-4 with ideal components and where the switch is repetitively operated such that $t_{ON} = \frac{1}{4}t_{OFF}$. Assume that L is so large that the L/R time constant is much greater than the switching period.
   a) Carefully sketch the steady-state waveforms for v_o and i.
   b) Calculate the steady-state power delivered to the 10-Ω load resistor.

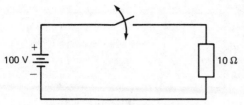

Fig. 1P-3. Simple switching circuit.

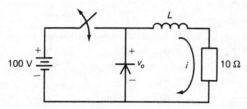

Fig. 1P-4. Switching circuit with L-R load and "coasting" diode.

10. Consider the switching circuit in Fig. 1P-4. Assume that the ideal switch is repetitively operated with the on-time equal to one-fourth of the off-time and that steady-state conditions exist. Use the steady-state principle for an inductor (see Section 1.5.3) to carefully sketch the waveforms for $v_o(t)$ and $Ri(t)$, assuming an ideal diode and that the L/R time constant is roughly the same as the switching period, $T = t_{ON} + t_{OFF}$. Also calculate the average current through the load resistor.

11. Consider the switching circuit shown in Fig. 1P-5. Assume that the switch is repetitively operated with $t_{ON} = \frac{1}{4}t_{OFF}$. What is the average current flowing through the resistor R in steady state?

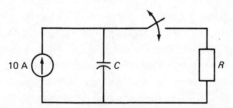

Fig. 1P-5. Current-source switching circuit.

12. Calculate the power delivered to the load resistor, assuming an ideal diode and resistive load in the circuit of Fig. 1P-6.

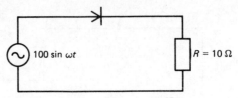

Fig. 1P-6. Half-wave rectifier.

13. Derive Eq. (1-30).
14. Use Eq. (1-24) to prove that
    a) waveforms with odd-function symmetry contain only sine terms in their Fourier series representations;
    b) waveforms with even-function symmetry contain only cosine terms in their Fourier series representations;
    c) waveforms with half-wave symmetry contain only odd harmonics in their Fourier series representations.
15. Determine the Fourier series representation for the square wave of Fig. 1-4, choosing the $\omega_1 t = 0$ axis at the beginning of the positive half-cycle.
16. Determine the Fourier series representations for each of the functions shown on Fig. 1P-7.

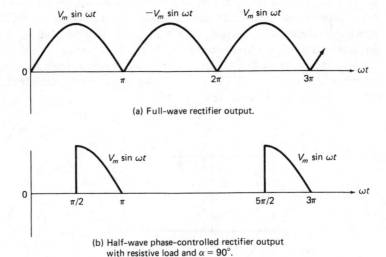

(a) Full-wave rectifier output.

(b) Half-wave phase-controlled rectifier output with resistive load and $\alpha = 90°$.

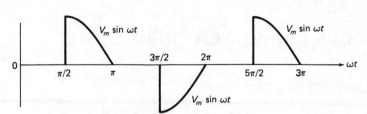

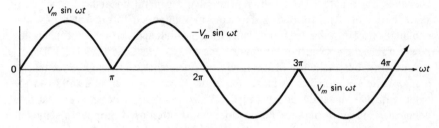

(c) Single-phase ac phase-control output
with with resistive load and $\alpha = 90°$.

(d) Cycloconverter-type output.

Fig. 1P-7. Waveforms for problem 16. (a) Full wave rectifier output; (b) half-wave phase-controlled rectifier output with resistive load and $\alpha = 90°$; (c) single-phase ac phase-control output with resistive load and $\alpha = 90°$; (d) cycloconverter-type output.

2
DIODES AND POWER TRANSISTORS

2.1. SEMICONDUCTOR PRINCIPLES

A knowledge of quantum mechanics and solid-state physics is necessary background for comprehending the details of the operation of semiconductor devices. However, for this text, it is sufficient to have a general understanding of transistor behavior. The following description is intended to present a physical explanation of bipolar transistor operation, which will provide the basis for power transistor application limitations and unique characteristics [1], [2].

Electron tubes make use of the flow of electric charges through a vacuum or a gas, while semiconductors make use of the flow of current in a solid. As the name indicates, a semiconductor material has a conductivity between that of a conductor and an insulator. The materials most often used in modern semiconductor devices are germanium and silicon. Most early power transistors were germanium devices, since its higher electrical conductivity provided low voltage drops at high currents. However, silicon has now displaced germanium in almost all applications. Silicon is more suitable for high-power devices because of its higher temperature capability and because modern processing techniques yield more economical devices with higher voltage capability.

Semiconductor materials have a crystal structure. In the particle model for this type of structure, the outer orbit or valence electrons are considered to be tightly bound to the corresponding electrons of adjacent atoms in electron-pair or covalent bonds. At low temperatures and without externally applied electric fields, such a structure has a negligible number of free electrons and thus is a poor conductor. One way to increase the conductivity is to add a high enough energy to break the electron-pair bonds and produce free electrons for electrical conduction. This can be done by applying very high temperatures or strong electric fields.

Another way to obtain free electrons is to "dope" the semiconductor with small amounts of "impurities" having a different number of valence electrons. The density of the impurities is very low, so that the doped material contains a sparse distribution of impurity atoms replacing silicon atoms throughout the crystal lattice. Each impurity atom is considered to be completely surrounded by silicon atoms. When the impurity atoms have more valence electrons than the semiconductor atoms, extra electrons are introduced into the lattice, and these extra electrons cannot form bonds with adjacent atoms since no other valence electrons are available. Such impurities are called "donors," since they add electrons to the lattice. These excess or extra electrons are held only very loosely by their atoms. A small excitation will break them away, permitting them to move through the material and resulting in greater conduction.

When impurity atoms called "acceptors" having fewer valence electrons are added to the semiconductor material, some of the valence electrons in the semiconductor atoms will have no adjacent electron with which to form electron-pair bonds. As a result, vacancies or "holes" exist in the crystal-lattice structure. An electron from an adjacent electron-pair bond may receive enough energy to break its bond and move through the lattice to fill a hole. As with the presence of excess electrons, the existence of holes makes it easier for electrons to flow in the semiconductor material, thereby increasing conductivity. The hole in the crystal structure is equivalent to a positive electrical charge because it represents the absence of an electron. It should be noted that the entire doped crystal, with either an excess of electrons or electron vacancies due to the doping, has no net positive or negative charge since each atom has no net charge. The doped semiconductor material with an excess of electrons is called *n*-type, while the material including acceptor impurities is called *p*-type.

2.2. DIODE BEHAVIOR

When *n*-type and *p*-type materials are joined together, a *p-n* junction is formed. Some of the relatively free electrons in the *n*-type material diffuse across the junction and recombine with holes in the *p*-type region. Similarly, some of the holes in the *p*-region diffuse across the junction and recombine with electrons in the *n*-type material. As a result of these effects, a slight negative charge is produced on the *p*-side and a slight positive charge on the *n*-side of the *p-n* junction. This produces a potential gradient or energy barrier that prevents further diffusion of charge carriers across the junction. The formation of the potential gradient prevents total interaction between the two types of materials and thus preserves their differences.

When an external potential is connected across a *p-n* junction, the amount of current flow is determined by the polarity of the applied voltage. With a positive voltage on the *n*-side, free electrons in the *n*-type material are attracted to the positive terminal of the external voltage source and, at the same time, holes from the *p*-region are attracted to the negative terminal of the external voltage. Then a space-charge or depletion region is formed at the junction. There is negligible carrier flow across the junction, and the junction is said to be reverse-biased.

When the positive terminal of an external voltage source is connected to the *p*-side of a *p-n* junction, electrons in the *n*-material are attracted across the junction toward the positive terminal of the external voltage. Electrons near the positive terminal of the external voltage break their electron-pair bonds and enter the external circuit, forming new holes in the *p*-region. Thus there is continuous electron flow across the junction from the *n*-region to the *p*-material so long as the external voltage is applied. In this case, the junction is said to be forward-biased.

The operation just described is that of a conventional semiconductor diode which contains a single *p-n* junction. Symbols for the diode and its *i-v* characteristic are shown in Fig. 2-1. In the forward direction, current rises rapidly with applied voltage. Within the normal operating range, the reverse current is very low even at high reverse voltage. Excessive reverse voltage can cause excessive currents, resulting in high temperatures and damage to the device because of additional carrier flow caused by the high-intensity electric fields produced from large reverse voltages. The forward current must be limited by the external circuit to a value that will not overheat the junction.

2.3. BIPOLAR JUNCTION TRANSISTORS

2.3.1. Operation

In this explanation of a bipolar junction transistor, an *npn* device is considered, since it involves the motion of electrons, which is a somewhat simpler concept than the motion of holes. Consider such a transistor biased as shown in Fig. 2-2. The excess carriers in each region are indicated by the encircled negative signs for electrons and the encircled positive signs for holes. It is also assumed that

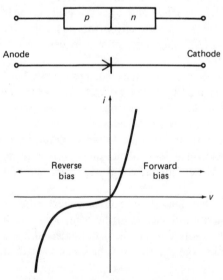

Fig. 2-1. Diode symbol and characteristic.

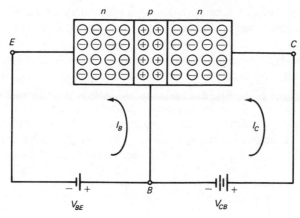

Fig. 2-2. *npn* transistor.

the excess carrier concentration in the emitter is considerably greater than in the base. Therefore, there are many more electrons available for conduction per unit volume in the emitter than there are holes available for conduction in the base. This means that the primary conduction will result from electron flow.

Because semiconductors contain two groups of independent charge carriers, which have charges of opposite sign, it is possible by adding electrical, optical, or thermal energy to the material to increase almost equally the concentrations of both holes and electrons without causing significant deviations from electrical neutrality. The carrier concentrations increase as a consequence of the breaking of valence bonds in the semiconductor. Thus, the mobile charge-carrier concentration can be increased without creating the enormous internal electric fields and currents normally associated with the deviations from neutrality that would accompany such increases in single-carrier conductors (e.g., metals). This fundamental difference between metals and semiconductors is directly responsible for the operation of most semiconductor junction devices. Processes which cause the carrier concentrations to increase above their equilibrium values are referred to as carrier injection mechanisms, and the added carriers are called excess carriers. [3]

Two mechanisms of conduction may occur in a semiconductor. The first is referred to as "drift" current. This is similar to the mechanism by which current flows in a conventional conductor. The charge carriers are moved by the force of an applied electric field. This type of current flow is governed by the relationship

$$J = \sigma \mathscr{E} \qquad (2\text{-}1)$$

where J is the current density in amperes per square centimeter, σ is the conductivity in ohms per centimeter, and $\mathscr{E}$ is the electric field intensity in volts per centimeter.

The second type of current flow in semiconductor devices is a diffusion current. Diffusion occurs whenever there is a difference in concentration of carriers in adjacent regions of a crystal. The diffusion current is directly proportional to the negative of the concentration gradient, the cross-sectional area of the device, and the charge of each carrier. However, the current from electron diffusion is in a direction opposite to the electron flow. Thus, the corresponding expressions for electron and hole diffusion currents are as follows:

$$I_n = qD_n A \frac{dn}{dx} \tag{2-2}$$

$$I_p = -qD_p A \frac{dp}{dx} \tag{2-3}$$

The terms D_n and D_p are diffusion constants for electrons and holes, respectively; A is the cross-sectional area, q is the magnitude of the charge on an electron, while dn/dx and dp/dx are the gradients for electrons and holes, respectively.

When a bias voltage is applied from base to emitter as indicated in Fig. 2-2, most of this voltage appears across the base-emitter junction. Excess electrons near this junction in the emitter are then forced across the junction by the action of the electric field produced by the base-emitter bias. This is referred to as electron injection from the emitter to the base. Once these electrons are in the base region, they now diffuse across the base. When they approach the collector-base junction, again they are forced across this junction by the electric field resulting from the collector-base voltage. In a good *npn* transistor, almost all of the electrons which are injected from the emitter into the base also are swept into the collector. Thus, approximately the same carrier flow, and current flow, results across each junction. Since the emitter-base junction is forward-biased and the collector-base is reverse biased with a much larger voltage, there is considerable power gain in the device. Some of the electrons injected into the base recombine with atoms in this part of the material, which results in a base current flow. The time it takes the carriers to diffuse across the base region is one of the basic limitations on the switching speeds for the transistor. The highest mobility carriers and narrowest base regions improve switching time but also generally reduce the voltage capability. The cross-sectional area determines the collector-current capability, and one of the major difficulties is to obtain complete uniformity in the material when large area junctions are fabricated.

2.3.2. Circuit Models

Although there are numerous circuit models for transistors, the low-frequency and large signal-switching representations are of principal importance to the solid-state power circuit designer.

The discussion in this section considers models for *pnp* transistors. The models for the *npn* transistor are fundamentally the same, although sign changes are required, as mentioned later, in certain of the mathematical relationships.

One of the most common low-frequency transistor equivalent circuits is given in Fig. 2-3 [4]. Each of the elements in this equivalent circuit can be related quite directly to the physical transistor. The current generator $\alpha' i_e$ represents the portion of the emitter current which flows across the collector junction; the resistor r_e' is the forward-biased diode resistance of the emitter-base diode; the resistor r_b' is added to represent the ohmic resistance between the base lead and the center of the base region; the conduction caused by minority carrier flow from the collector to the base is represented by resistor r_c'; and the feedback voltage generator accounts for the base-width modulation effect. It is sometimes more convenient to transform the circuit of Fig. 2-3 to that shown in Fig. 2-4. This can be accomplished in a straightforward manner mathematically. The circuit model of Fig. 2-4 is the conventional T-equivalent circuit which is quite widely used as a low-frequency transistor representation.

On most specification sheets, *h*-parameters are given to characterize the transistor in its linear region of operation. These are a hybrid set of parameters determined from the terminal variables in a two-port representation of the transistor. Figure 2-5 shows the general circuit representation. The defining relationships for the *h*-parameters are the following:

$$v_1 = h_{11}i_1 + h_{12}v_2 \tag{2-4}$$

$$i_2 = h_{21}i_1 + h_{22}v_2 \tag{2-5}$$

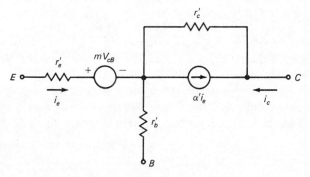

Fig. 2-3. Low-frequency transistor equivalent circuit.

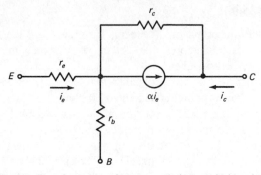

Fig. 2-4. Transformed low-frequency transistor equivalent circuit.

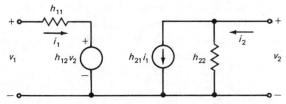

Fig. 2-5. General two-port representation.

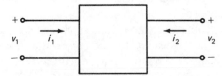

Fig. 2-6. Equivalent circuit in terms of h-parameters.

The equivalent circuit in terms of the hybrid parameters is shown in Fig. 2-6. It should be noted that it is possible to write the relations representing the two-port network of Fig. 2-5 in a number of different ways. The coefficients of the variables in the relations may be impedances; admittances; current transfer ratios; voltage transfer ratios; or a combination of these, which is the case for the h-parameter set.

The h-parameters may be calculated in terms of T-equivalent circuit parameters using the circuit of Fig. 2-4 and relationships corresponding to those given in Eqs. (2-4) and (2-5). These relationships are written in the following form:

$$v_{eb} = h_{11}i_e + h_{12}v_{cb} \tag{2-6}$$

$$i_c = h_{21}i_e + h_{22}v_{cb} \tag{2-7}$$

For this common base arrangement, the h-parameters are also often written as follows:

$$v_{eb} = h_{ib}i_e + h_{rb}v_{cb} \tag{2-8}$$

$$i_c = h_{fb}i_e + h_{ob}v_{cb} \tag{2-9}$$

The Ebers-Moll model is widely used as a nonlinear representation of the transistor for large-signal switching operation. The equations for this model of a *pnp* transistor are [5]:

$$I_E = I_{ES}[\varepsilon^{(-qV_{EB}/kT)} - 1] - \alpha_R I_{CS}[\varepsilon^{(qV_{CB}/kT)} - 1] \tag{2-10}$$

$$I_C = -\alpha_F I_{ES}[\varepsilon^{(qV_{EB}/kT)} - 1] + I_{CS}[\varepsilon^{(qV_{CB}/kT)} - 1] \tag{2-11}$$

The corresponding equations for an *npn* transistor are given below for reference purposes to indicate the sign changes involved:

$$I_E = -I_{ES}[\varepsilon^{(-qV_{EB}/kT)} - 1] + \alpha_R I_{CS}[\varepsilon^{(-qV_{CB}/kT)} - 1] \tag{2-12}$$

$$I_C = \alpha_F I_{ES}[\varepsilon^{(-qV_{EB}/kT)} - 1] - I_{CS}[\varepsilon^{(-qV_{CB}/kT)} - 1] \tag{2-13}$$

The four parameters α_F, α_R, I_{ES}, and I_{CS} are related by the following expression:

$$\alpha_R I_{CS} = \alpha_F I_{ES} \tag{2-14}$$

Figure 2-7 shows the circuit representation of the Ebers-Moll model for a *pnp* transistor. The first term on the right-hand side of Eq. (2-10) represents the emitter-base diode current which flows as a result of minority carrier injection into the base. For the *pnp* transistor, the minority carriers in the base region are holes which are injected from the *p*-emitter region because of the influence of the emitter-base bias voltage.

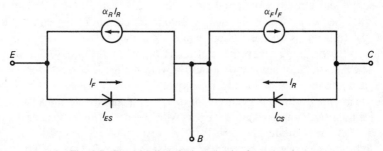

Fig. 2-7. Ebers-Moll equivalent circuit of *pnp* transistor.

The first term on the right-hand side of Eq. (2-11) gives the fraction of the carrier flow across the emitter-base junction which also flows across the base-collector junction. The parameter called the "forward" short-circuit gain, α_F, ranges usually from 0.9 to 1.00. It is slightly less than unity for two reasons

(i) A small fraction of the holes injected into the base at the emitter recombine in the base or on its surface before reaching the collector.
(ii) Not all of the emitter current is a result of hole injection into the base. Even in a well designed transistor, some of the emitter current is the result of electrons which are injected into the emitter by the base [6].

The other two terms on the right-hand sides of Eqs. (2-10) and (2-11) have a similar interpretation for reverse operation of the transistor. The current gain α_R is called the reverse short-circuit gain and it is a measure of the fraction of the holes injected from the collector into the base which are transported across the base region and then flow across the emitter-base junction. The parameter α_R may vary from near unity to much less than unity, depending on the transistor structure.

The Ebers-Moll model may also be transformed to a form where the current sources are controlled by the terminal currents rather than by diode currents. For this purpose, new saturation currents are defined as follows:

$$I_{EO} = I_{ES}(1 - \alpha_F \alpha_R) \qquad (2\text{-}15)$$

$$I_{CO} = I_{CS}(1 - \alpha_F \alpha_R) \qquad (2\text{-}16)$$

These definitions are used with Eqs. (2-10) and (2-11) to give

$$I_E = -\alpha_R I_C + I_{EO}[e^{(qV_{EB}/kT)} - 1] \qquad (2\text{-}17)$$

$$I_C = -\alpha_F I_E + I_{CO}[e^{(qV_{CB}/kT)} - 1] \qquad (2\text{-}18)$$

The equivalent circuit corresponding to Eqs. (2-17) and (2-18) is shown in Fig. 2-8. The current I_{EO} is the reverse saturation current of the emitter diode when the collector is open-circuited, while I_{CO} is the reverse saturation current of the collector diode when the emitter is open. In the previous model of Fig. 2-7, the saturation currents I_{ES} and I_{CS} were the diode currents for the collector and emitter short-circuited, respectively.

For many applications it is sufficient to use an approximate model which is less accurate but simpler than the Ebers-Moll model. This is the ideal diode model. The ideal diode has zero forward resistance when conducting and infinite back resistance when it is reverse-biased. Figure 2-9 shows the equivalent circuit for this approximate model.

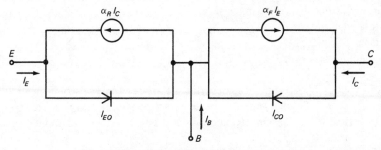

Fig. 2-8. Transformed Ebers-Moll equivalent circuit.

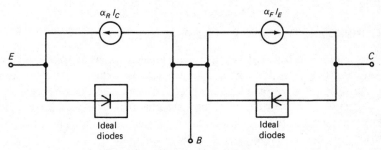

Fig. 2-9. Ideal diode equivalent-circuit model.

Each of the models previously discussed may be used to represent the operation of the transistor in large signal switching applications. The regions of operation possible are listed below.

1. The cutoff region — both junctions reverse-biased.
2. The normal region — emitter forward-biased and collector reverse-biased.
3. The inverse region — collector forward-biased and emitter reverse-biased. (The operation is essentially the same as in the normal region, except that the roles of the collector and emitter are interchanged.)
4. The saturation region — both junctions forward-biased.

2.3.3. Fabrication

The following devices illustrate several fabrication techniques used for modern power transistors.

Alloy-junction transistors
Diffused-junction transistors
Hometaxial-base transistors
Microalloy diffused-junction transistors

Grown diffused-junction transistors
Mesa transistors
Planar transistors
Epitaxial transistors

References [7] through [13] describe details of the construction of these devices and include characteristic comparisons. Reference [14] is an excellent survey article on processing technology and characteristics of modern power semiconductor devices.

Numerous types of power transistors are available from a variety of manufacturers. Most silicon units are *npn* transistors. However, there are a number of complementary *pnp* silicon power transistors presently available.

2.3.4. Static and Dynamic Characteristics [15]–[17].

2.3.4.1. Collector current rating. The greatest value of collector current permissible generally is determined by either the minimum current gain desired or by the fusing current of internal connecting wires and bonding techniques. Both continuous and peak current ratings are specified. The peak current rating is usually not greater than twice the continuous current rating. Thus, the power transistor does not have a large surge current capability.

A regenerative condition known as "thermal runaway" also places a limit on the greatest permissible value of collector current. Thermal runaway occurs when the thermally generated carrier concentration approaches the impurity concentration. This condition can occur as a result of the average temperature over a large area of the junction, or it may be produced in a small area by localized hot spots. In either case, the thermally generated carriers predominate. This results in an increase in carrier flow which causes a higher junction temperature due to the internal heating, and this produces more thermally generated carriers, further increasing the carrier flow in a regenerative fashion.

2.3.4.2. Collector voltage rating. The maximum voltage rating is determined by either a breakdown voltage level or a punch-through voltage. The transistor breakdown voltage is the value of voltage between external terminals of the device above which the crystal structure changes and the current begins to increase rapidly. After the initial breakdown for collector-to-emitter voltage, the voltage decreases with increasing collector current until another breakdown occurs at a lower voltage. The voltage then remains essentially constant for increasing collector current. This minimum collector-to-emitter breakdown voltage is called the sustaining voltage.

The voltage can be limited below the breakdown levels as a result of the punch-through effect. In this situation, the collector-base voltage may become high enough so that the depletion layer at this junction expands through the base region and then essentially contacts the emitter junction.

2.3.4.3. Saturated voltage drop. Transistor saturation is defined as the point above which an increase in the base current gives no further significant increase in its collector current. The saturated voltage drop is the voltage appearing across the external terminals when saturation occurs. Usually the rating sheet will specify both $V_{CE(SAT)}$ and $V_{BE(SAT)}$. These values are a function of the specific operating conditions, i.e., the collector current, the base drive current, and the junction temperature. For an ideal transistor switch, $V_{CE(SAT)}$ is zero. However, in practical high-current switching transistors, the collector-emitter saturation voltage ranges from millivolts to several volts. Device dissipation for dc or low-frequency switching operation is determined primarily by the saturated voltage drop, $V_{CE(SAT)}$, multiplied by the on-state collector current.

2.3.4.4. Current gain. Two types of current gain are of major interest in the application of high-current power transistors — the large signal or dc current gain h_{FE} and the small signal or ac gain, h_{fe}. Most high-current applications use the common emitter circuit. The small signal or ac gain defines the change in the collector current for a corresponding small change in the base drive current:

$$h_{fe} = \frac{\Delta I_c}{\Delta I_b}\bigg|_{I_c} \qquad (2\text{-}19)$$

This value is a function of the steady state or quiescent value of the collector current, the frequency of the variation in the base drive current, and the operating junction temperature. For linear application, such as low-frequency and audio power amplifiers, the small signal ac current gain is a most significant parameter. In switching applications, the dc gain

$$h_{FE} = \frac{I_c}{I_b} \qquad (2\text{-}20)$$

is of most interest. This is the ratio of the dc value of the collector current to the corresponding dc value of the base current. The dc gain defines the amount of base drive required to limit the saturated voltage drop to the desired level. In high-frequency switching applications, the base drive current is not a constant dc value. Rather it is switched from a rather large on-state value to either zero or some negative value during the off-state. However, the minimum on-interval usually is considerably larger than the transient switching times of the device so that the dc current gain, implying the current gain after transient effects are over, is still the most significant gain parameter. In high-frequency switching applications the base drive currents in both the on-state and the off-state are critical. Switching operation with minimum power dissipation during the switching is achieved by the optimum combination of on-state and off-state base drive currents.

2.3.4.5. Power dissipation. A semiconductor device is heated by the electrical energy dissipated in it [18]. The maximum power rating for a transistor is given to assure that the temperature in all parts of the device is limited below a value that will result in detrimental changes in device characteristics.

The source of the heat is the energy dissipated in the active portions of the transistor. It is a rather complex heat-flow problem to determine the dynamics of the heat removal and thus the transient junction temperature. The term *junction temperature* is used to imply the temperature of the active semiconductor device, including one junction in the case of diodes, but two junctions for transistors. The device temperature is assumed to be the same throughout the entire semiconductor chip on which the device is fabricated. Case temperature is, as the name implies, the temperature of the housing in which the device is enclosed. This is different from the junction temperature because of the thermal impedance between the semiconductor chip and the case — due primarily to the imperfect thermal conductivity of the heat-flow paths through internal connections and bonding elements. The heat-sink temperature is also assumed the same at all points on the heat sink. This is not exactly true because of thermal resistance in the sink, but it is a good practical assumption for a well-designed heat sink. Finally, the ambient air temperature determines the amount of heat flow from the sink to the ambient for a given cooling means and a specified heat-sink configuration. The heat flow from the sink to the ambient results in an additional temperature drop, and thus an equivalent thermal resistance from the sink to the ambient air.

Transistor dissipation is produced by on-state loss, off-state loss, and switching loss. In general, leakage currents are so low that the off-state losses can be considered negligible. Three modes of transistor operation must be considered when calculating device dissipation. The first mode is very low-frequency switching or dc operation. In this case, the switching losses are negligible and thus the total device dissipation is caused by on-state losses.

The second mode of operation which is important from the transistor dissipation standpoint is intermediate-frequency switching. In this mode, it is assumed that both switching and on-state losses are significant. In addition, the thermal time constant of the transistor may be such that there is significant transient variation in the device junction temperature over each operating switching cycle. During the on-interval, the junction temperature rises in accordance with the equivalent thermal capacity of the entire device package and the thermal resistance, from the device junction to the ambient. In a like manner, the junction temperature falls during the off-interval as determined by the transistor thermal time constant.

For the third mode of operation of importance in the calculation of power dissipation, the transistor is operating in a high-frequency switching mode. More precisely, the thermal time constant of the device is assumed to be much longer than the period of the switching cycle. In this situation, the transistor dissipation is again caused by both switching losses and on-state losses. However, the variation in device junction temperature over an operating switching cycle is very small, and the switching losses may be the dominant losses.

2.3.4.6. Switching times ([19], [20]).
Figure 2-10 shows a simplified test circuit used for measuring transistor switching times. The important waveforms during switching are illustrated in Fig. 2-11. When a positive going input voltage pulse is applied from base to emitter, the base current immediately rises. However, the collector current does not increase until some time later. This delay time is defined as the time interval from the application of the input voltage pulse until the collector current has increased to 10 percent of its on-state value. Delay time t_d is caused by the time it takes to charge the transistor junction capacitances to new values. The time it takes for the collector current to rise from 10 to 90 percent of its final on-state value, the rise time t_r, is determined primarily by the collector junction capacitance and the transit time of charge carriers in the base region.

When the base-emitter input voltage pulse is removed, the collector current does not immediately start to decrease. This storage time, t_s, is the time required for the excess charge in the collector and base regions to recombine and produce the charge distribution that exists when the transistor is just ready to come out of saturation. The rate of charge decay is determined by the minority carrier lifetime in the collector and base regions, the amount of reverse turn-off current i_{B2}, and the magnitude of the turn-on current i_{B1}. Storage time is measured as the time interval beginning at the instant when the input voltage is removed and ending when the collector current has fallen to 90 percent of its on-state value.

The transistor fall time t_f is roughly the reverse process of the rise time. It is determined primarily by collector junction capacitance. However, it is also

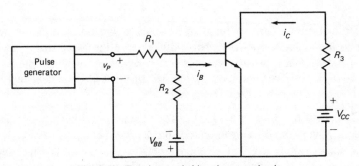

Fig. 2-10. Transistor switching-time test circuit.

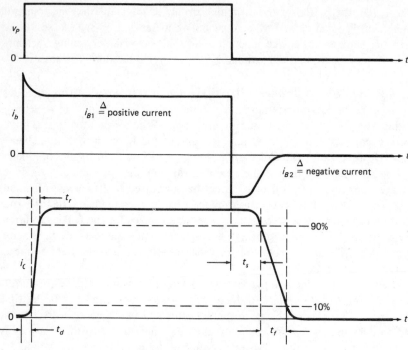

Fig. 2-11. Transistor switching.

reduced by increasing the negative base current i_{B2}. As indicated in Fig. 2-11, the fall time is the interval from 90 to 10 percent of the on-state current.

The turn-on and turn-off times are defined as follows:

$$t_{ON} = t_d + t_r \qquad (2\text{-}21)$$

$$t_{OFF} = t_s + t_f \qquad (2\text{-}22)$$

It is extremely important to minimize these switching times so that transistor switching losses are minimized. This will increase the power-handling capability of a particular transistor when used in a repetitive switching application. Application data sheets provided by the device manufacturer describe circuit techniques for reducing switching times.

2.3.4.7. Sustaining voltage [17]. The sustaining voltage is a means of defining a minimum collector-to-emitter breakdown voltage. Transistor breakdown voltages define the voltage values between two specified terminals at which the current begins to rise rather rapidly. The voltage then remains rela-

tively constant over a wide range of current. Breakdown voltages may be specified with the third terminal open, shorted or biased in either the forward or
reverse direction. For example, the minimum collector-to-emitter breakdown
voltage with the base open is the sustaining voltage designated $V_{CEO(SUS)}$.

A test circuit for measuring the sustaining voltage $V_{CEO(SUS)}$ is shown in
Fig. 2-12. An approximate oscilloscope waveform is shown in Fig. 2-13. The
transistor is initially in the on-state. Then the switch is opened to remove the
base drive current. As the collector current begins to fall, an $L(di/dt)$ voltage is

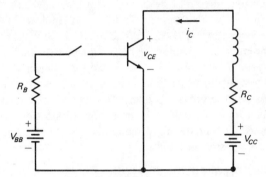

Fig. 2-12. Sustaining-voltage test circuit.

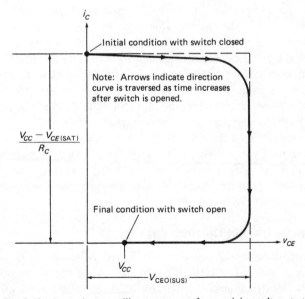

Fig. 2-13. Approximate oscilloscope pattern for sustaining-voltage test.

induced across the inductor to oppose this reduction in current. If the transistor were to abruptly turn off, the inductor voltage would approach infinity. However, when it reaches the sustaining-voltage breakdown level, the collector-emitter voltage remains approximately constant, and this implies a linear rate of reduction in collector current. After the base switch has been opened for some time, the transistor will be in the off-state with only a very small leakage-current flow and with approximately the collector supply voltage, V_{CC}, appearing across the collector-emitter. For the practical measurement of sustaining voltage it is possible to use a storage scope and a single switch opening. However, the oscilloscope trace may be barely visible during the rapidly changing portions of the $i_C - v_{CE}$ characteristic. Thus, it is often desirable to repetitively operate the base switch of Fig. 2-12 at some relatively low duty cycle. If this is done, a continuous scope display of the characteristic in Fig. 2-13 can be observed. It is very important to minimize lead inductances in the test circuit, and it is also necessary to use a base switch like a mercury relay contact that opens and closes without contact bounce. The size of the inductor and the on-state current must be consistent with the energy-dissipation capability of the transistor under test. Essentially all of the initial stored energy in the inductor is dissipated in the transistor during the test.

2.3.4.8. Second breakdown [21]–[26].
This is a potentially destructive phenomenon that occurs in all bipolar transistors. When the transistor is switched off, the emitter-base junction is reverse-biased and a lateral or transverse electric field is produced in the base region which causes current crowding under the center of the emitter. This constricts the current flow to a small portion of the base region, producing localized hot spots in the transistor pellet. If the energy in these hot spots is sufficient, the excessive localized heating can destroy the transistor. Forward-biased second breakdown also can occur during turn-on of the transistor. However, the localized current densities are greater during reverse-biased second breakdown. Thus, the energy capability of the transistor is much lower when it is reverse-biased.

The second-breakdown limiting effect is most severe during switching transients when relatively high voltages and high currents occur simultaneously for finite time intervals. Power transistors with narrow base regions, which are required to achieve high-frequency switching characteristics, are more severely limited by second breakdown.

2.3.5. Example Design Calculations

Generally, extensive testing is required before a particular transistor, with an appropriate heat sink, is selected for a production design. This must include comprehensive electrical tests to determine maximum steady-state and transient device voltages and currents over the expected range of operating conditions, and considerable thermal testing to establish worst-case device junction temperatures. Throughout the component selection process, there should be frequent contact with the transistor manufacturer. The published specification data accommodate worst-case situations and provide general guidance to the user. A particular device will be used to the limits of its capability, which is essential for economic production designs, only if the knowledge possessed by both the device application engineer and the circuit designer is put to full use.

In this section, several design examples are discussed to illustrate important design considerations. A Westinghouse D62T transistor is used since it is one example of currently available high-power switching transistors. (Appendix II includes the specifications for the D62T, and Appendix III contains application information on this type of transistor.)

2.3.5.1. Low frequency.
The simplest application of a power transistor is for static switching, i.e., solid-state relays, contactors, and circuit breakers. Choppers or buck regulators operating at low frequency also offer a similar design problem. In these situations, the switching losses are very small because the time it takes the transistor to switch from "on" to "off," or vice versa, is short compared to the switching period.

Consider the simple circuit of Fig. 2-14(a) operated as a static switch. It would be rather uneconomical to use the D62T with a 125-V supply and in the static switching mode, since it is a 400- to-500-V transistor capable of switching operation in the 100-kHz range. However, this relatively low-voltage static switch illustrates the simplest thermal design problem. Generally, the off-state losses are quite small. Thus, they are neglected in this and subsequent design examples. The base drive losses are not negligible. However, they are considerably smaller than the on-state losses in this example, and for simplicity, they also will be neglected. Since the maximum collector-emitter voltage is only one-fourth of the highest $V_{CEO(SUS)}$ rating, second-breakdown limitations will not be considered. For the static switch, it is assumed that the transistor is switched on for a long period of time — at least several seconds and possibly much longer [see Fig. 2-14(b)]. In summary, the following conditions are assumed:

1. No second breakdown limitation
2. Negligible off-state loss
3. Negligible base drive loss
4. $V_{CE(SAT)} = 1.2$ V and $I_{B1} = 20$ A (Note: I_{B1} is the on-state base drive current.)

The typical collector-emitter saturation-voltage curves for the higher current version of the D62T (see Appendix II) indicate a saturation voltage of about 1.2 V with $I_{B1} = 20$ A and at a junction temperature of 150°C. This yields a

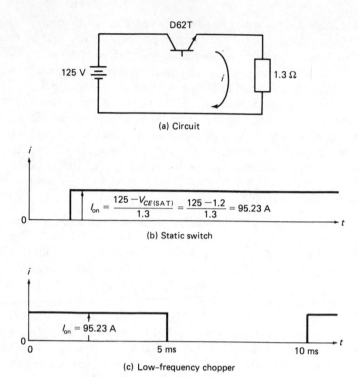

(a) Circuit

$$I_{on} = \frac{125 - V_{CE(SAT)}}{1.3} = \frac{125 - 1.2}{1.3} = 95.23 \text{ A}$$

(b) Static switch

$I_{on} = 95.23$ A

5 ms   10 ms

(c) Low-frequency chopper

Fig. 2-14. Static switch or low-frequency chopper.

continuous on-state power loss in the transistor of

$$P_T = \frac{125 - 1.2}{1.3} \times 1.2 = 114.28 \text{ W} \qquad (2\text{-}23)$$

The dc thermal resistance from junction to sink for double-sided cooling is 0.14°C/W (see Appendix II). Thus, the junction-to-sink temperature differential is

$$\Delta T_{JS} = R_{\theta JS} \times P_T = 0.14 \times 114.28 = 16°C \qquad (2\text{-}24)$$

Figure 2P-2 indicates that with two of the smaller heat sinks, curve (b), for double-sided cooling, the sink-to-ambient temperature rise would be approximately 80°C with 114.28-W dissipation in the transistor. This means that an ambient temperature of 54°C would be permissible without exceeding 150°C junction temperature.

$$T_J = T_A + \Delta T_{JS} + \Delta T_{SA} = 54° + 16° + 80° = 150°C \qquad (2\text{-}25)$$

A variation of the static switch application, in which switching losses are still negligible, is the low-frequency chopper. Consider the same circuit as in Fig. 2-14(a) with the same assumptions as in the previous example, but now operating as a chopper. Assume a chopper period of 10 ms and a 50 percent duty cycle. Again it would be rather uneconomical to use the D62T at such a low switching frequency, but this variation on the previous example will further illustrate a rather simple thermal design problem. Figure 2-14(c) shows the chopper current waveform. In this case, the transistor power loss is

$$P_T = V_{CE(SAT)} \times I_{ON} \times \frac{t_{ON}}{T} = 1.2 \times 95.23 \times \frac{1}{2} = 57.14 \text{ W} \qquad (2\text{-}26)$$

and the junction to sink "average" temperature differential is

$$\Delta T_{JS} = R_{\theta JS} \times P_T = 0.14 \times 57.14 = 8°C \qquad (2\text{-}27)$$

In this instance, the transistor junction temperature will increase during the on-interval and decrease during the off-interval, assuming that the thermal time constant of the transistor is in the same order as, or shorter than, the chopper period. This can be shown from transient thermal impedance data. A good discussion of transient thermal impedance is contained in references [27] and [28].

The calculation of the transient variation in junction temperature is accomplished as follows. Initially, assume that the transistor junction, case, and sink are all at the ambient air temperature. For this example, it will be assumed also that the transistor case and sink temperatures remain constant at the ambient air temperature. Of course, the case and heat-sink temperatures must rise after sufficient time has elapsed. However, the thermal mass of the case and the thermal mass of the heat sink are both much larger than that of the semiconductor chip. Thus, the time required for the case and sink temperatures to change is much longer than the time for the junction temperature to change. (Reference [27] includes a discussion of the effect of the heat sink on the transient thermal impedance characteristic.) The junction temperature rise is determined using the transient thermal impedance for the D62T (see Appendix II) with a sequence of step-inputs of transistor dissipation. A step-input of power equal to the on-state loss occurs at the beginning of each switching period, and an equal but negative step-input of power takes place at the end of each on-interval. Figure 2-15 illustrates the step-inputs of dissipation and the initial transient variation in the junction temperature, which is calculated as follows:

$$\Delta T_{JC}(1 \text{ ms}) = Z_{\theta JC}(1 \text{ ms}) \times 114.28 \text{ W} = .003°C/W \times 114.28 \text{ W} = 0.34°C \qquad (2\text{-}28)$$

$$\Delta T_{JC}(3 \text{ ms}) = .0045 \times 114.28 = 0.51°C \qquad (2\text{-}29)$$

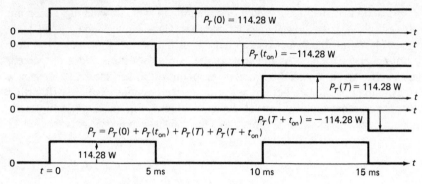

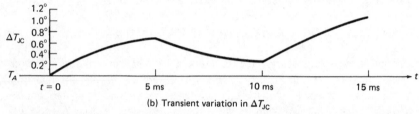

(a) Equivalent step-function representation of transistor dissipation.

(b) Transient variation in ΔT_{JC}

Fig. 2-15. Initial variation in ΔT_{JC}. (a) Equivalent step-function representation of transistor dissipation; (b) transient variation in ΔT_{JC}.

$$\Delta T_{\text{JC}}(5 \text{ ms}) = .006 \times 114.28 = 0.69°C \qquad (2\text{-}30)$$

$$\Delta T_{\text{JC}}(8 \text{ ms}) = [Z_{\theta\text{JC}}(8 \text{ ms}) - Z_{\theta\text{JC}}(3 \text{ ms})] \times 114.28$$

$$= (.0075 - .0045) \times 114.28 = 0.34°C \qquad (2.31)$$

$$\Delta T_{\text{JC}}(10 \text{ ms}) = [Z_{\theta\text{JC}}(10 \text{ ms}) - Z_{\theta\text{JC}}(5 \text{ ms})] \times 114.28$$

$$= (.0085 - .006) \times 114.28 = 0.29°C \qquad (2\text{-}32)$$

$$\Delta T_{\text{JC}}(13 \text{ ms}) = [Z_{\theta\text{JC}}(13 \text{ ms}) - Z_{\theta\text{JC}}(8 \text{ ms}) + Z_{\theta\text{JC}}(3 \text{ ms})] \times 114.28$$

$$= (.01 - .0075 + .0045) \times 114.28 = 0.8°C \qquad (2\text{-}33)$$

$$\Delta T_{\text{JC}}(15 \text{ ms}) = [Z_{\theta\text{JC}}(15 \text{ ms}) - Z_{\theta\text{JC}}(10 \text{ ms}) + Z_{\theta\text{JC}}(5 \text{ ms})] \times 114.28$$

$$= (.012 - .0085 + .006) \times 114.28 = 1.09°C \qquad (2\text{-}34)$$

It is clear from Fig. 2-15(b) that the equivalent thermal time constant of the D62T is considerably longer than 10 ms. As indicated by Eq. (2-27), the steady-state "average" ΔT_{JS} is 8°C. However, ΔT_{JC} has only risen to 1.09°C 15 ms after the chopper begins operating.

The steady-state junction temperature may be obtained by continuing the process shown in Fig. 2-15 for many cycles of the chopper frequency. However, another approximate method can be used which does not require nearly as many calculations. For this approach, it is assumed that the chopper has been operating for many cycles. An initial junction temperature is calculated using the average power dissipation and the dc thermal resistance. Then two consecutive power pulses are applied, with the actual duty cycle, and the junction temperature is calculated at the end of the second power pulse. This gives a worst-case or maximum junction temperature which is within a few percent of the value calculated using a sequence of power pulses, as in Fig. 2-15, but continuing until the steady-state junction temperature is reached. Figure 2-16 illustrates the equivalent representation of the transistor power dissipation, and the calculations are as follows:

$$\Delta T_{JC}(t_{ON} + T) = V_{CE(SAT)} \times I_{ON} \times \frac{t_{ON}}{T} \times R_{\theta JC}$$

$$+ [V_{CE(SAT)} \times I_{ON} - P_{T,AVE}]Z_{\theta JC}(t_{ON} + T)$$

$$- V_{CE(SAT)} \times I_{ON} \times Z_{\theta JC}(T) + V_{CE(SAT)} \times I_{ON} \times Z_{\theta JC}(t_{ON})$$

$$= V_{CE(SAT)} \times I_{ON}\left[\frac{t_{ON}}{T} \times R_{\theta JC} + \left\{1 - \frac{t_{ON}}{T}\right\} \times Z_{\theta JC}(t_{ON} + T)\right.$$

$$\left. - Z_{\theta JC}(T) + Z_{\theta JC}(t_{ON})\right]$$

$$= 114.28\left[\frac{1}{2} \times .09 + \frac{1}{2} \times .012 - .0085 + .006\right]$$

$$= 114.28 \times 0.485 = 5.54°C \tag{2.35}$$

The total junction-to-sink temperature rise must include the effect of the thermal resistance from the case to the sink. As previously stated, this has not been considered in Eq. (2-35) because the thermal time constant of the case is assumed to be considerably larger than the junction time constant. The following expression is used to calculate the approximate steady-state junction-to-sink

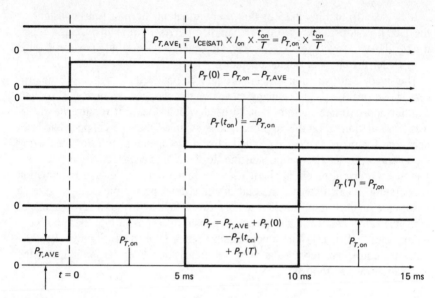

Fig. 2-16. Equivalent representation of transistor power dissipation for steady-state ΔT_{JC} calculation.

temperature rise:

$$\Delta T_{JS} = \Delta T_{JC} + \Delta T_{CS} = 5.54°C + P_{T,AVE} \times R_{\theta CS}$$

$$= 5.54°C + \frac{114.28}{2} \text{ W} \times .05°C/W = 5.54°C + 2.86°C = 8.4°C$$

$$(2\text{-}36)$$

This is reasonably close to the value calculated in Eq. (2-27). It is somewhat higher because of the transient variation in temperature which occurs during the on-interval. The highest instantaneous temperature occurs at the end of each on-interval.

2.3.5.2. Intermediate frequency. Figure 2-17 illustrates a simple single-phase bridge transistor inverter. For this design example, the following assumptions are made:

1. Square wave operation at 5000 Hz
2. Resistive load = 2 Ω
3. Supply voltage E = 250 V
4. Negligible off-state loss

5. Negligible base drive loss
6. $V_{CE(SAT)} = 1.2$ V and $I_{B1} = 20$ A
7. Linear switching during both turn-on and turn-off
8. The rise time, t_r, and fall time, t_f, of the transistor current during turn-on and turn-off, respectively, are each equal to 1 μs; i.e., $t_r = t_f = 1$ μs.

In this case, the frequency of switching is high enough so that there is negligible change in the junction temperature during an on-interval or an off-interval. This means that the transient thermal impedances for the very short on and off intervals are very small compared with the dc thermal resistance. Thus, it is permissible to use the steady-state or dc thermal resistance in the calculation of junction temperature rise.

$$\Delta T_{JS} = (\text{turn-on switching loss} + \text{on-loss} + \text{turn-off switching loss}) \times R_{\theta JS}$$
$$(2\text{-}37)$$

The switching loss expressions are as given in Chapter 1, Section 1.4.1. Thus,

$$\Delta T_{JS} = \left(\frac{E \times I_{ON}}{6} \times \frac{t_r}{T} + V_{CE(SAT)} \times I_{ON} \times \frac{t_{ON}}{T} + \frac{E \times I_{ON}}{6} \times \frac{t_f}{T} \right) \times R_{\theta JS}$$

$$= \left(\frac{250 \times 123.8}{6} \times \frac{2\ \mu s}{200\ \mu s} + 1.2 \times 123.8 \times \frac{98\ \mu s}{200\ \mu s} \right) \times 0.14°C/W$$

$$= (51.58\ W + 72.97\ W) \times 0.14°C/W = 17.41°C \qquad (2\text{-}38)$$

2.3.5.3. High frequency. For this final example, the following conditions are assumed for the single-phase inverter of Fig. 2-17:

1. Square wave operation at 50 kHz
2. Resistive load = 2 Ω
3. Supply voltage $E = 250$ V
4. Negligible off-state loss
5. $V_{CE(SAT)} = 1.2$ V and $I_{B1} = 20$ A and $V_{BE(SAT)} = 1.6$ V
6. Linear resistive switching during turn-on and linear inductive switching during turn-off
7. The rise time, t_r, and the fall time, t_f, of the transistor current during turn-on and turn-off, respectively, are each equal to 1 μs.

Again it is permissible to use the steady-state or dc thermal resistance in the calculation of junction temperature rise. The switching loss during turn-on can

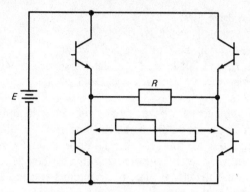

Fig. 2-17. Single-phase bridge inverter.

be calculated in the same way as in the previous example. However, the turn-off situation is different. It is assumed that there is sufficient inductance in the collector current so that a high $L(di/dt)$-voltage is produced across this inductance when the collector current is turned off quickly. This induced voltage causes the collector-emitter voltage to rise immediately to the sustaining-voltage breakdown level. For simplicity, it is assumed that $V_{\text{CEO(SUS)}}$ is equal to the collector supply voltage. Generally, $V_{\text{CEO(SUS)}}$ should be somewhat greater than the supply voltage. Figure 2-18 illustrates the assumed turn-off characteristic. The turn-off switching loss is determined as follows:

$$P_T\Big|_{\text{turn-off}} = \frac{1}{t_f} \int_0^{t_f} [E]\left[I_{\text{ON}}\left(1 - \frac{t}{t_f}\right)\right] dt = \frac{E \times I_{\text{ON}}}{2} \qquad (2\text{-}39)$$

Thus,

$$\Delta T_{\text{JS}} = \left(\frac{E \times I_{\text{ON}}}{6} \times \frac{t_r}{T} + V_{\text{CE(SAT)}} \times I_{\text{ON}} \times \frac{t_{\text{ON}}}{T}\right.$$

$$\left. + V_{\text{BE(SAT)}} \times I_{B1} \times \frac{t_{\text{ON}}}{T} + \frac{E \times I_{\text{ON}}}{2} \times \frac{t_f}{T}\right) \times R_{\theta\text{JS}}$$

$$= \left(\frac{250 \times 123.8}{6} \times \frac{1}{20} + 1.2 \times 123.8 \times \frac{8}{20}\right.$$

$$\left. + 1.6 \times 20 \times \frac{8}{20} + \frac{250 \times 123.8}{2} \times \frac{1}{20}\right) \times 0.14$$

$$= (257.92 + 59.42 + 12.8 + 773.75) \times 0.14$$

$$= 1103.89 \text{ W} \times 0.14°\text{C/W} = 154.54°\text{C} \qquad (2\text{-}40)$$

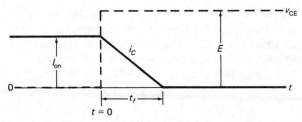

Fig. 2-18. Inductive turn-off.

The base drive losses were included in this calculation for illustrative purposes, although they are quite small. Comparing Eqs. (2-38) and (2-40), the average transistor losses are much greater for the high-frequency (50 kHz) operation. Also, of course, the turn-off switching losses increase due to the inductive switching assumption. Equation (2-39) shows that the turn-off switching losses now are three times as great as the turn-on switching losses. As a result, the junction-to-sink temperature rise is now greater than permissible. Assuming a perfect heat sink, the ambient temperature would need to be limited to about 45°C to stay within the 200°C maximum allowable junction temperature.

As a final illustration, the previous calculations are repeated for the same conditions except that the frequency is reduced to 25 kHz. Then, from Eq. (2-40),

$$\Delta T_{\text{JS}} = \left(\frac{250 \times 123.8}{6} \times \frac{1}{40} + 1.2 \times 123.8 \times \frac{18}{40} \right.$$

$$\left. + 1.6 \times 20 \times \frac{18}{40} + \frac{250 \times 123.8}{2} \times \frac{1}{40} \right) \times 0.14$$

$$= (128.96 + 66.85 + 14.4 + 386.88) \times 0.14$$

$$= 597.09 \text{ W} \times 0.14°\text{C/W} = 83.59°\text{C} \tag{2-41}$$

With double-side cooling, two of the largest heat sinks, (i) on Fig. 2P-2, would result in a ΔT_{SA} of approximately 70°C with 597.09 W transistor dissipation. This means that an ambient air temperature of about 45°C would be required to limit the maximum junction temperature to less than the 200°C allowable.

$$T_J = T_A + \Delta T_{\text{JS}} + \Delta T_{\text{SA}} = 45 + 83.59 + 70 = 198.59°\text{C} \tag{2-42}$$

2.3.5.3. Second-breakdown limitations. In the design examples presented, a procedure is not given for assuring that second-breakdown failures are avoided. One technique which is used to define forward-bias safe operation is as follows. The maximum permissible transistor dissipation is determined using the procedure of Fig. 2-16, but including all losses in Eq. (2-40). This means that ΔT_{JS} is set equal to the maximum allowable value based on the maximum operating junction temperature and an assumed case or heat-sink temperature. Then the maximum permissible transistor dissipation is calculated. A constant power line is drawn on the forward-bias SOA set of curves, and a second-breakdown curve is drawn from this line, roughly paralleling the nearest second-breakdown curve on the data sheet. Safe operation is assured if the peak instantaneous power does not exceed that defined by the calculated constant power line and graphically determined second-breakdown limiting region. Reference [26] discusses this approach in some detail. The problem with the approach is that it gives quite conservative results. In addition, forward-bias second-breakdown failure is not as likely to result as is reverse-bias second-breakdown failure [30]. Generally, the reverse-bias SOA is rectangular for high-power transistors, at the limits of the continuous current rating, and only slightly below the $V_{CEO(SUS)}$ rating. However, it is very important to minimize switching time and to limit the instantaneous power during switching. Snubber circuits across the transistor reduce the probability of second-breakdown failure during turn-off. The circuit design engineer must work closely with the device manufacturer to make certain that second-breakdown failures are avoided, without excessively conservative application of devices.

2.4. ADDITIONAL REMARKS

There are a number of additional topics which are important in the application of diodes and transistors. Diodes are often used in series and parallel, which requires special design considerations. Power transistors also are used in parallel, and this requires even more careful design to assure current sharing transiently during turn-on and turn-off as well as during the on-intervals.

In general, each manufacturer has its own approach to presenting design information on device data sheets. Thus, it is necessary for the design engineer to become familiar with the specification information for the particular device to be used. Also, it is extremely important to have considerable interaction between circuit-design and device-application engineers. The second-breakdown limitations of power transistors generally are not well defined on published data sheets. Thus, this is one area where it is particularly important to consult device manufacturers.

REFERENCES

1. *RCA Transistor, Thyristor and Diode Manual.* Somerville, NJ: RCA Corporation, 1971, pp. 3–8.

2. Comer, Donald T. *Large Signal Transistor Circuits*. Englewood Cliffs, NJ: Prentice-Hall, Inc., 1967.

3. Searle, C. L., et al. *Elementary Circuit Properties of Transistors*. New York: John Wiley & Sons, 1964, pp. 6–7.

4. Comer, Donald T. Op. cit., pp. 21–26, 37–39.

5. Searle, C. L. Op. cit., pp. 37–54.

6. Ibid., pp. 21–24.

7. Comer, Donald T. Op. cit., pp. 28–34.

8. *RCA Transistor, Thyristor and Diode Manual*. Op. cit., pp. 12–15.

9. *RCA Designer's Handbook Solid State Power Circuits*. Somerville, NJ: RCA Corporation, 1971, pp. 99–112.

10. Roehr, William D., et al. *Switching Transistor Handbook*. Phoenix: Motorola, Inc., 1970, pp. 14–23.

11. Hower, P. L. Optimum Design of Power Transistor Switches. IEEE Transactions on Electron Devices ED-20: April 1973, pp. 426–435.

12. *Power Data Book*. Mountain View, CA: Fairchild Camera and Instrument Corporation, 1976, chap. 1.

13. Owyang, K., and Shafer, P. A New Power Transistor Structure for Improved Switching Performance. Washington, DC: IEEE International Electron Devices Meeting, December 1978. (Also available as General Electric Application Note 660.37.)

14. Pelly, Brian R. Power Semiconductor Devices — A Status Review. Orlando: IEEE/IAS International Semiconductor Power Converter Conference Record, 1982, pp. 1–19.

15. *RCA Designer's Handbook Solid State Power Circuits*. Op. cit., pp. 113–149.

16. Hunter, Lloyd P., ed. *Handbook of Semiconductor Electronics*, 3d ed. New York: McGraw-Hill Book Company, 1970, pp. 19-8 through 19-15.

17. *RCA Transistor, Thyristor and Diode Manual*. Op. cit., pp. 17–37.

18. *RCA Designer's Handbook: Solid State Power Circuits*. Op. cit., pp. 123–126.

19. Ibid., pp. 175–182.

20. *RCA Transistor, Thyristor and Diode Manual*. Op. cit., pp. 30–35.

21. *RCA Designer's Handbook: Solid State Power Circuits*. Op. cit., pp. 126–134.

22. *The Power Semiconductor Data Book for Design Engineers*, 1st ed. Dallas: Texas Instruments, Inc., pp. 11-4 through 11-17.

23. *Power Data Book*. Mountain View, CA: Fairchild Camera and Instrument Corporation, 1976, chap. 2.

24. Ghandi, Sarab K. *Semiconductor Power Devices*. New York: John Wiley & Sons, 1977, pp. 172–183.

25. Hawkins, H. R. *Mechanisms of Secondary Breakdown in Power Transistors*. Kokomo: Delco Electronics, May 12, 1977.

26. Bennett, Wilfred P., and Kumbatovic, Robert A. Power and Energy Limitations of Bipolar Transistors Imposed by Thermal-Mode and Current-Mode Second-Breakdown Mechanisms. IEEE Transactions on Electron Devices ED-28: October 1981, pp. 1154–1162.

27. *SCR Manual*, 6th ed. Auburn, NY: General Electric, 1979, pp. 37–42.

28. Gutzwiller, F. W., and Sylvan, T. P. Power Semiconductors Under Transient and Intermittent Loads. AIEE Transactions, Part I, Communications and Electronics, January 1961, pp. 699–706.

29. *Westinghouse Silicon Power Transistor Handbook*. Youngwood, PA: Westinghouse Electric, 1967, pp. 2-19 through 2-22.

30. Hower, P. L., and Tarneja, K. S. The Influence of Circuit and Device Parameters on the Switching Performance of Power Transistors. Power Conversion International 6, July/August 1980, pp. 10–22.

PROBLEMS

1. Derive the relations which express the parameters of Fig. 2-4 *in terms of those* in Fig. 2-3. (Hint: Write the z-parameter equations for each circuit and use the relations obtained from equating the z-parameters.)
2. Write equations corresponding to Eqs. (2-4) and (2-5) for z-parameters and y-parameters.
3. Find the h-parameters for the circuit in Fig. 2P-1. (Hint:

$$h_{11} = \frac{v_1}{i_1}\bigg|_{v_2=0} \; ; \qquad h_{12} = \frac{v_1}{v_2}\bigg|_{i_1=0} \; ; \qquad h_{21} = \frac{i_2}{i_1}\bigg|_{v_2=0} \; ; \qquad h_{22} = \frac{i_2}{v_2}\bigg|_{i_1=0}$$

Apply an arbitrary value of the denominator variable in each case with v_2 or i_1 equal to zero as indicated; then calculate the numerator variable and determine the required ratio.)

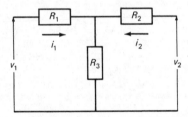

Fig. 2P-1. Resistive T-circuit.

4. Derive the expressions for h_{ib}, h_{rb}, h_{fb}, and h_{ob} in terms of the parameters of Fig. 2-4.
5. Derive the expression for h_{fe} in terms of parameters of Fig. 2-4. (Note: $v_{be} = h_{ie}i_b + h_{re}v_{ce}$; $i_c = h_{fe}i_b + h_{oe}v_{ce}$.)
6. Write the approximate expressions resulting from Eqs. (2-10) and (2-11) for each of the four regions of operation for a switching transistor.
7. Derive Eqs. (2-17) and (2-18) from Eqs. (2-10), (2-11), (2-15), and (2-16).
8. Sketch the characteristic curves I_E vs. V_{EB} with I_C as a parameter, and I_C vs. V_{CB} with I_E as a parameter, for the approximate model in Fig. 2-9.
9. Consider a transistor carrying a dc current of 20 A with a $V_{CE(SAT)}$ of 3 V and with a thermal resistance from junction-to-sink of 1°C/W. Select a suitable heat sink from those shown in Fig. 2P-2, assuming a maximum permissible junction temperature of 150°C, an ambient air temperature of 70°C, and single-side cooling.

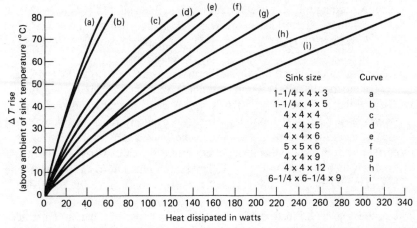

Fig. 2P-2. Standard heat sink ratings* for natural convection — aluminum extrusion. *Zinc-chromate converse coating. (From Westinghouse. Used with permission)

10. Consider the simple transistor switching circuit with linear, repetitive switching as shown in Fig. 2P-3.

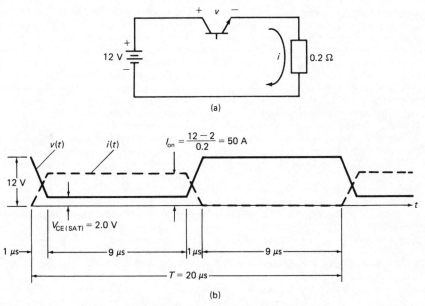

Fig. 2P-3. Switching-circuit and switching parameters.

   a) Assuming the off-state current is negligible and that $V_{CE(SAT)} = 2.0$ V, calculate the total average watts dissipated in the transistor. (Hint: Use Eqs. (1-1), (1-3), and (1-4) of Chapter 1, suitably modified to include the effect of $V_{CE(SAT)}$ on switching loss.)

   b) Assuming a junction-to-case thermal resistance of 1.0°C/W and a case temperature of 90°C, calculate the maximum junction temperature.

11. Assume that a Westinghouse D62T transistor with double-sided cooling is switched on at $t = 0$ supplying 200 A to a purely resistive circuit. Also assume negligible turn-on switching time with a step application of base current such that the collector-emitter current is a 200-A step at $t = 0$, following which $V_{CE(SAT)} = 2.0$ V. Calculate the junction-to-case temperature rise at $t = 1$ ms, $t = 100$ ms, and $t = 10$ seconds after the transistor is switched on.

12. Use the approach illustrated in Fig. 2-16 to calculate the maximum steady-state ΔT_{JC} for the example in Section 2.3.5.2.

3
THYRISTORS

3.1. TERMINAL PROPERTIES

The most widely used member of the thyristor family of semiconductor devices is the silicon-controlled rectifier, which is a reverse-blocking triode thyristor, hereafter referred to as the SCR. Since the SCR first became available in late 1957, it has become of great importance in the power control and conversion technology. Devices are produced now using 100-mm diameter silicon wafers. These thyristors may have ratings of 4000 V and 2500 A average [1]. The SCR has low voltage drop when it is on, negligible leakage when it is off, extremely fast switching speeds with low-power gate drive, and the reliability associated with all solid-state semiconductor devices. Thus, it is an excellent approximation of the ideal power switch.

The thyristor name implies a bistable semiconductor device which switches to the on-state by a p-n-p-n regenerative process. A number of thyristor devices have been produced in addition to the SCR. These include triacs, light-activated silicon-controlled rectifiers, asymmetrical SCRs, reverse conducting thyristors, bidirectional diode thyristors, reverse-blocking diode thyristors, static induction thyristors, and gate turn-off thyristors ([2]–[8]). The triac is widely used in ac phase-control applications. Gate turn-off thyristors show considerable promise for use in forced commutated inverters and choppers, since commutating circuits can be eliminated or significantly simplified. Each of the other thyristors presently available has considerable application potential, but the SCR is still the most widely used member of the thyristor family. For this reason, and also because it illustrates the behavior of thyristor devices, this chapter is devoted to the operation and characteristics of the SCR.

Figure 3-1 indicates the circuit symbol for the SCR along with a schematic representation of its structure. As its name implies, the SCR operates like a diode rectifier which can be controlled. The device always blocks current flow in the reverse direction, as does a plain diode. In the forward direction, the SCR normally also blocks, but it can be turned on in this direction either by exceeding the forward breakover voltage or by the application of a gate firing signal. When the device is triggered on by either of these methods, it remains on even though the gate signal is removed. Its forward voltage drops to a very low value when it is on — somewhat greater than the forward drop of a diode with the same current rating.

The i-v characteristic of the SCR is indicated in Fig. 3-2.

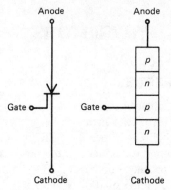

Fig. 3-1. SCR circuit symbol and schematic representation.

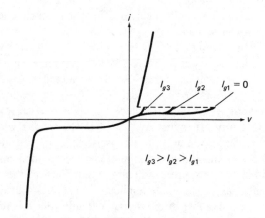

Fig. 3-2. SCR i-v characteristic.

It is also possible to trigger an SCR on by imparting other forms of energy to the device such as light, heat, or radiation. Some light-controlled devices are available, but all present devices for high-power application use gate control.

Since the SCR is a bistable switching device, it is quite analogous to a gas thyratron, while the transistor is more analogous to the vacuum triode. Both the transistor and the vacuum triode also can be used in switching-type circuits. However, the bistable characteristic of the SCR means that only a short-duration gate pulse is required for firing—then the gate signal can be removed so that the average gate power required to control an SCR is very small. For devices presently available, the power gain of a power transistor operated as a switch is in the range of 10^4, whereas SCR devices have power gains in excess of 10^6 with dc gating and much greater power gains with pulse triggering.

3.2. TWO TRANSISTOR ANALOG

A relatively straightforward representation of the SCR that demonstrates its regenerative switching action is shown in Fig. 3-3. In this analog model, the SCR is considered as two transistors. The center n-p regions of the SCR are common to both the p-n-p and n-p-n transistors. For the p-n-p transistor, the base current is given by

$$I_{B1} = I_A - I_{C1} = I_A - \alpha_1 I_A - I_{CO1} = (1 - \alpha_1)I_A - I_{CO1} \qquad (3\text{-}1)$$

The latter expressions are obtained by using the transistor equation, Eq. (2-18), considering junction $J2$ a reverse-biased collector junction. The collector current for the n-p-n transistor can be given by the expression

$$I_{C2} = \alpha_2 I_K + I_{CO2} \qquad (3\text{-}2)$$

Combining Eqs. (3-1) and (3-2) yields

$$(1 - \alpha_1)I_A - I_{CO1} = \alpha_2 I_K + I_{CO2} \qquad (3\text{-}3)$$

and using the additional relation,

$$I_K = I_A + I_g \qquad (3\text{-}4)$$

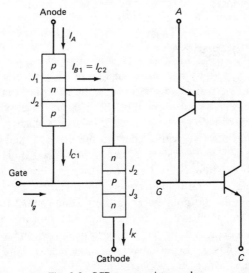

Fig. 3-3. SCR two-transistor analog.

the following expressions are obtained.

$$(1 - \alpha_1)I_A - I_{CO1} = \alpha_2(I_A + I_g) + I_{CO2}$$

$$I_A(1 - \alpha_1 - \alpha_2) = \alpha_2 I_g + I_{CO1} + I_{CO2}$$

$$I_A = \frac{\alpha_2 I_g + I_{CO1} + I_{CO2}}{1 - \alpha_1 - \alpha_2} \tag{3-5}$$

It is also possible to obtain this same expression by equating the base-drive current, I_g, plus the collector current, I_{C1}, to the base current of the n-p-n transistor.

The parameter α_1 is the fraction of the holes injected from the p-type emitter (the anode) of the p-n-p transistor, which are collected at its collector-base junction, junction $J2$. For the n-p-n transistor, α_2 is the fraction of the electrons injected from the n-type emitter (the cathode), which are collected at its collector-base junction, again junction $J2$.

Equation (3-5) indicates that if $\alpha_1 + \alpha_2$ becomes equal to unity, the anode current can increase without bound, which results in a type of regenerative action. Since the current gains of transistors are affected by changes in the emitter current, if the gate drive raises the emitter current to the point where the sum of the alphas is unity, the SCR switching action will occur. This switching action may also be induced by other means. The application of light, a sufficient increase in device temperature, or raising the anode voltage can also cause enough increase in the emitter current to produce the situation where $\alpha_1 + \alpha_2 = 1$.

3.3. PHYSICAL THEORY [9]

For a somewhat deeper physical understanding of the regenerative action in the SCR, the hole and electron concentrations can be examined during device turn-on. A schematic representation of the SCR for this discussion is shown in Fig. 3-4. If a forward voltage less than the breakover voltage is slowly applied, the anode becomes positive with respect to the cathode, and junction $J2$ becomes reverse-biased, while $J1$ and $J3$ are lightly forward-biased. The device remains in the off condition. Electrons near junction $J2$ move toward the positively based anode, leaving behind donor impurities that are stripped of electrons; and in a similar fashion, holes in the $p2$ side of $J2$ move toward the cathode, leaving uncompensated acceptor impurities on the right side of $J2$. The result is that a depletion region composed of donors and acceptors uncompensated by mobile charge carriers develops in the region of $J2$, creating a high electric field which sustains the applied voltage.

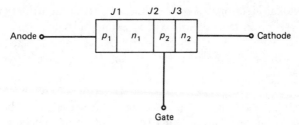

Fig. 3-4. SCR schematic representation.

The equilibrium carrier concentrations in the regions of the device of Fig. 3-4 are assumed to be as follows:

$$p1 \qquad 10^{19} \text{ acceptors/cm}^3$$

$$n1 \qquad 10^{14} \text{ donors/cm}^3$$

$$p2 \qquad 10^{16} \text{ acceptors/cm}^3$$

$$n2 \qquad 10^{19} \text{ donors/cm}^3$$

These relative concentrations are typical of those required for desirable operation of practical devices.

When the forward voltage is raised to the breakover voltage level, the following events occur:

1. The increased forward voltage produces a larger forward bias on $J1$. This means that more holes are injected from region $p1$ into $n1$. Electron injection from right to left across $J1$ may be neglected.
2. The holes injected from $p1$ to $n1$ diffuse across region $n1$ and are then swept across $J2$ because of the large electric field at $J2$. This junction had been reverse-biased by nearly the full anode-cathode applied voltage.
3. The holes swept across $J2$ increase the hole density in region $p2$. This produces a momentary build-up of positive charge in region $p2$ which raises the effective forward bias on $J3$.
4. The greater forward bias on $J3$ causes more electrons to be injected from $n2$ to $p2$ as in the normal forward conducting diode. Because of the relative carrier concentrations, hole injection from $p2$ to $n2$ is negligible compared to the increased electron injection from right to left across $J3$.
5. The electrons injected into region $p2$ diffuse across this region and are swept into the $n1$ region by the electric field across $J2$.

6. The additional electrons swept into the $n1$ region momentarily raise the negative charge in this region, which results in greater forward bias on $J1$. This increased forward bias causes more holes to be injected from $p1$ to $n1$. This is the beginning of event (1), and thus a regenerative switching action occurs.

The regenerative switching action continues until the anode-cathode voltage across the device drops to in the order of one volt. This low value of voltage results because the electric field across $J2$ continues to sweep holes from $n1$ to $p2$ and electrons from $p2$ to $n1$ until a sufficient number of the uncompensated donors and acceptors in the space charge region of $J2$ become compensated to forward bias this junction. In the steady-state forward conducting condition, the major carrier-flow involves holes moving from left to right across $J1$ and electrons from right to left across $J3$. Recombination in the relatively long $n1$ and $p2$ regions must occur to provide the necessary hole and electron currents across $J2$ to sustain the carrier flow at the terminals of the device.

Although the turn-on action has been discussed assuming it was initiated by increasing the anode-cathode voltage to the breakover level, a similar action is initiated when a gate current is applied. The gate circuit increases the forward bias on junction $J3$, producing a sequence of events starting with event (4), which again produces the regenerative switching action.

The anode-cathode voltage across the device in the forward conducting condition, neglecting the bulk resistance of the material, is the sum of the voltages on the three junctions:

$$V_{AK} = V_{J1} + V_{J2} + V_{J3} \qquad (3\text{-}6)$$

Since the "back-injection" on $J2$ is not as great as the injection across the other two junctions, the forward voltage of the SCR when it is on is somewhat greater than the forward drop of a similar diode conducting the same current. For example, the corresponding diode and SCR drops might be typically 0.8 V and 1.1 V, respectively.

An interesting feature of the "turned-on" condition of the SCR is that the middle junction $J2$ is forward biased while the current flowing through it is a reverse current with respect to this junction [10]. Thus, junction $J2$ in this condition is similar to the collector-base junction in the transistor of Fig. 2-2 when it is in saturation, to the thermoelectric generator junction, or to the diode junction during its reverse recovery time. However, the "generator" behavior of the diode junction in a rectifying device is a short duration carrier recombination or clean-up period, while the "generator" character of junction $J2$ is maintained continuously when the SCR is in its forward conducting state. The forward power loss is less than the sum of the theoretical losses in junctions $J1$ and $J3$ and may be calculated from the terminal voltage drop and current as in a simple rectifier element.

This has been a brief discussion of the physical theory of the SCR. For a more complete discussion of this subject, the reader may refer to a number of semiconductor texts and articles including references [11] through [13].

3.4. STATIC AND DYNAMIC CHARACTERISTICS

3.4.1. Turn-On

In most circuit applications, the SCR is turned on by the application of a gate firing signal. A dc voltage can be applied to the gate. The value of the voltage must be sufficient to trigger the device under the desired conditions. Usually SCR specification sheets will indicate the minimum gate voltage to fire all units of the specified type and the minimum gate current necessary from the gating source. There is considerable variation in the gating requirements from unit to unit of a given type. In addition, the gating requirements usually are much less at high operating temperatures. It is necessary to design the gate signal source such that it is sufficient to fire all units under worst-case conditions without exceeding the allowable maximum gate dissipation.

In many cases it is preferable to apply a pulse to the gate. With a steep front on the pulse, the precise time of triggering can be determined accurately. A fast rise-time pulse is also required to achieve the di/dt rating capability. The pulse voltage magnitude and corresponding pulse power can be considerably greater than the minimum voltage to fire the SCR without the average gating power exceeding the maximum allowable gate dissipation. The minimum pulse width and the actual pulse voltage and current required to fire a given SCR may vary considerably, depending upon the operating temperature and the anode circuit [14]. With a large inductance in the anode-cathode load circuit, it may be necessary to use a gating pulse width considerably longer than that needed with a resistive load. In many applications it is desirable to use "continuous gating," which means applying a gating pulse during the entire possible conducting interval.

When SCR devices are operated in series or parallel it is most important to fire all of the units simultaneously. Usually some sort of pulse gating is used in these applications. The pulses to all gates must be applied almost exactly at the same instant, which means all pulses must be applied at the same time with very steep wavefronts. The gating circuits should be designed so that the combined effects of rise time of the gating pulses and delays between pulses are less than one-tenth the turn-on time.

3.4.2. Rate of Rise of Forward Current di/dt

Particularly in the higher rating devices, the forward current may be concentrated in a small part of the cross-sectional area of the SCR when it is first turned on. This can cause localized excessive dissipation and damage to the unit if the rate of rise of forward current is too great. Usually, a higher gate current will cause a larger part of the SCR cross section to carry current faster after the application of the gate signal. "Hard firing" techniques have been employed to increase di/dt capability. With these techniques, a large short-duration pulse followed by a smaller pulse is applied to the gate. This is done to start conduction in the maximum amount of the SCR cross section as fast as possible. However, devices employing some type of "regenerative gate" are now available with quite high di/dt ratings, several hundred amperes per microsecond. Multiple-gate devices have also been fabricated to provide devices with higher di/dt capability.

3.4.3. Rate of Rise of Forward Voltage dv/dt

The SCR may also turn on without the application of a gate signal if forward voltage is applied at too rapid a rate. An explanation for this phenomenon is that the internal capacitances of the SCR junctions can draw enough charging current to cause triggering [15]. In some applications, it is necessary to add circuit elements to limit the rate of rise of forward voltage. However, the "shorted emitter" type construction has resulted in a substantial increase in the dv/dt capability. Units are available with several hundred volts per microsecond capability. In essence, this technique provides a shunting path in parallel with the gate-cathode junction. In fact, the addition of an external resistance from gate to cathode increases the dv/dt capability, but this unfortunately also increases the gate drive requirements.

3.4.4. Power Dissipation

The power dissipated in an SCR results from the following:

1. Switching losses
2. Forward conduction loss
3. Blocking loss
4. Triggering losses

Forward conduction loss is often the major source of junction heating. However, at higher frequency operation or when high di/dt capability is necessary, the switching and triggering losses may become significant. The RMS current flowing through the SCR must be calculated for the worst-case conducting period and current waveform to be sure that it is within the RMS current rating specification. This will make sure that the forward conduction losses are within the device capability. Curves are generally available from the manufacturer to include the effects of switching losses for higher frequency operation or the effects on the device losses of unusual current wave shapes. Information is also given on the specification data sheets to enable the user to select appropriate heat sinks to increase the forward dissipation without exceeding the allowable maximum junction temperature, based on the thermal characteristics of the device.

One of the advantageous features of the SCR is its relatively high surge-current rating. This may approach ten times the RMS forward-current rating. The surge is assumed to be nonrepetitive and of short duration, often expressed as a peak sine wave current for a half-cycle of 60 Hz. This rating also assumes that the device is already on, rather than turned on into a fault which would often result in exceeding the allowable di/dt.

3.4.5. Forward and Reverse Voltage

Usually a repetitive peak reverse voltage and repetitive peak forward blocking voltage rating are specified and these are equal. These are defined for given gate conditions. If the reverse voltage is exceeded, a device failure generally results. When the peak repetitive forward voltage is exceeded, the device will usually be turned on without damage. However, the blocking voltage varies considerably with junction temperature so that in some cases an SCR is damaged if large forward voltages are applied at low temperature to cause forward breakover [16]. Avalanche-controlled rectifiers are available from some manufacturers. These devices are self-protecting to overvoltage in either direction, provided that the I^2t capability for the resulting surges is not exceeded.

3.4.6. Turn-Off

Although the forward current through a device may drop to zero almost instantly if the anode circuit is opened, the turn-off time is the time required for the SCR to regain its forward-blocking capability after forward conduction. This time is in the 10- to 50-μs range for the devices in which this is one of the important design parameters. Immediately after an SCR has been conducting, charge carriers exist in the regions of the device, which are at energy levels enabling them to participate in conduction. Until complete "recombination" of these carriers has taken place, the SCR may turn on when a forward voltage is reapplied.

The turn-off time is often referred to as the commutating time. Particularly in inverters and dc-to-dc converters, one of the major circuit-design problems is to provide the necessary circuits to accomplish commutation. Complete commutation in such a circuit involves several steps, including stopping the flow of current through a previously conducting SCR, preventing forward voltage from being reapplied during the turn-off time, and starting the flow of current in the next SCR to conduct [17].

The turn-off time in a given device is a function of the forward current prior to turn-off, the magnitude of reverse voltage applied to stop the flow of current, the amount and rate of reapplied forward voltage, and the junction temperature. In repetitive-switching-type SCR circuits such as inverters and many dc-to-dc converters, the SCR turn-off time sets the upper limit on the allowable switching frequency.

3.4.7. Reverse Recovery [18]

Immediately after forward conduction through an SCR, there may be significant reverse current flow. For this to occur, the external circuit must act in a way to attempt to force negative current. This can result due to the application of negative anode-cathode voltage or due to resonant circuit action which attempts to produce a negative half-cycle of current through the SCR. The time interval immediately after forward conduction during which reverse current flows is called the reverse recovery time. This is the time required in a diode or SCR for the blocking junction to regain its reverse blocking capability. During the major part of the reverse recovery time, the junction which is to block remains forward-biased even though a rather large reverse current may flow. Depending upon circuit conditions and device characteristics, the peak reverse recovery current may approach the magnitude of the forward current which had been flowing, and the recovery time may range from in the order of 1 μs to 100 μs. The reverse recovery current and reverse recovery time increase as the forward current and rate of reduction of forward current to zero are increased.

The reverse recovery time is the time interval required for charges which had been participating in conduction to be removed so that they no longer can result in current flow. On most diode or SCR data sheets, recovery time information is presented in terms of stored charge.

3.5. SPECIFICATION SHEETS

The General Electric C434/C435 specification sheet is included in Appendix IV. This is a high-speed SCR that is designed for power switching at frequencies considerably higher than 60 Hz. Several specific applications of the C434 are considered to illustrate the use of SCR specification sheets in practical design.

3.5.1. Phase-Controlled Rectifier (1000 Hz)

Two C434 devices are used in a single-phase-controlled rectifier, as shown in Fig. 3-5, operating from a single-phase 1000-Hz sinusoidal power supply.

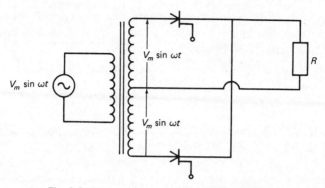

Fig. 3-5. Single-phase, center-tap-controlled rectifier.

Since even with an efficient heat sink there will be considerable temperature rise from ambient air to the SCR stud, a case temperature of 90°C is assumed. Figure 2 of Appendix IV gives a maximum allowable peak on-state current for each SCR of approximately 1000 A. This assumes a pulse base width of 500 μs, since the SCR current is a half sinewave at maximum output, when each SCR is gated at the positive-going zero crossing of its applied forward voltage. It should be noted that the curves on Figure 2 of the specification sheet are for very specific operating conditions — a switching voltage of 400 V, 2000-lb mounting force, double-side cooled, and an RC snubber of 0.25 μF in series with 5 Ω across the SCR. In addition, it is very important to have sufficient gate drive to minimize switching losses and to withstand high di/dt. The switching losses also are directly affected by the forward switching voltage and the maximum reverse voltage. If switching voltages greater than 400 V are encountered, the allowable peak current would be less than given on the curves of Figure 2 in Appendix IV. The dotted curves on Figure 2 indicate the reduction in maximum allowable peak on-state current when the maximum reverse voltage approaches 400 V.

Figure 5 of the specification sheet is used next to determine the SCR dissipation. For the 1000-A peak on-state current and a 500-μs pulse width, the dissipation is approximately 0.5 watt-second per pulse. Therefore, the average SCR dissipation is

$$W_{\text{SCR}} = 0.5 \text{ watt-second/pulse} \times f = 0.5 \times 1000 \text{ Hz} = 500 \text{ W} \qquad (3\text{-}7)$$

Thus, the heat sink would need to be capable of handling 500-W dissipation with the temperature differential allowed between the ambient and the 90°C stud temperature. This would require a very large extruded aluminum heat sink if using natural convection or forced-air cooling.

It is interesting to determine the RMS current through the SCR for the maximum allowable peak current. With 180° conduction and sinusoidal current, the relationship between the RMS and peak current is

$$I_e = \frac{I_m}{2} \tag{3-8}$$

Thus, the SCR RMS current is 700 A with 1400-A peak current. Generally, a greater RMS current is permissible at lower operating frequency where the switching losses are less significant. Also, a greater RMS current can flow through the SCR when the case temperature is low, since this means that more watts can be dissipated in the SCR without exceeding the maximum junction temperature. From Figure 1 of the C434/C435 specification sheet, the maximum allowable peak on-state current is approximately 1800 A for 500-μs pulse base width operation at 1000 Hz with a 65°C case temperature.

3.5.2. Phase-controlled Rectifier (5000 Hz)

Again assuming a 90°C case temperature, Figure 2 of Appendix IV gives a peak on-state current of about 250 A for a 100-μs pulse base width at 5000 Hz. These are the full-on conditions in the single-phase controlled rectifier of Fig. 3-5 operating from a 5000-Hz sinusoidal supply with a device reverse voltage approaching 400 V. Figure 5 of the specification sheet gives roughly 0.035 watt-second/pulse for 250-A peak and 100-μs pulse width. Thus, the average SCR dissipation is

$$W_{\text{SCR}} = 0.035 \text{ watt-second/pulse} \times f = 0.035 \times 5000 \text{ Hz} = 175 \text{ W} \tag{3-9}$$

It is interesting to note that the SCR average dissipation has been reduced to about one-third, while the peak on-state current is about one-fourth of the value at 1000 Hz. The peak on-state current must be reduced since the switching losses are so much greater at 5000 Hz than at 1000 Hz. However, one might expect the average SCR dissipation allowable to be the same at 1000 Hz and 5000 Hz. This is not the case, because at 5000 Hz the SCR is operating at such a high speed that only a part of the total junction cross section conducts during each on-interval.

3.5.3. SCR Chopper (1000 Hz)

Consider a C434 SCR used in the basic switching circuit of Fig. 1-1(a) under the following conditions:

Maximum case temperature = 90°C
Voltage E = 400 V
Rate-of-rise of on-state current = 100 A/μs
2000-lb mounting force
Double-side cooled
0.25-μF, 5-Ω RC snubber
Duty cycle = 50 percent (the on-time is equal to the off-time)
Reverse diode in parallel with the SCR ($V_R \leq$ 10 V)

It must be noted that the circuit of Fig. 1-1(a) would require a means for commutating the SCR, but this is not shown nor considered in this example. From Figure 7 of Appendix IV the peak on-state current allowable is about 500 A. Figure 10 gives approximately 0.45 watt-second/pulse. Thus, the SCR dissipation is

$$W_{SCR} = 0.45 \text{ watt-second/pulse} \times f = 0.45 \times 1000 = 450 \text{ W} \qquad (3\text{-}10)$$

For a rectangular current waveform with 50 percent duty cycle (that is, with the on-time equal to the off-time), the RMS and maximum currents are related as follows:

$$I_e = \sqrt{\frac{1}{2\pi} \int_0^\pi (I_m)^2 \, d(\omega t)} = \frac{I_m}{\sqrt{2}} \qquad (3\text{-}11)$$

Thus, the RMS current for the 500-A peak allowable on-state current is approximately 350 A. A lower value of RMS current is necessary for the chopper than for the 1000-Hz phase-controlled rectifier because the switching losses are greater with the more abrupt switchings in the chopper.

The C434/C435 specification sheet has been used for these examples because it illustrates one rather convenient rating philosophy. Another important example that can be handled with the sinewave data of this specification sheet is the case of commutating SCRs. The current waveform in such SCRs generally can be approximated by a sinusoidal pulse followed by an off-period many times longer than the on-interval. Transient thermal impedance data also can be used to obtain the SCR peak allowable current for pulse duty as described in references [19] and [20].

These examples have been included to illustrate some of the basic aspects of the use of SCR rating sheets. A number of additional important device-related considerations are involved in any practical design—including cooling, packaging, protection, optimum gate drive circuits, and techniques for operating devices in series or parallel. Generally, the best sources of design information, relative to optimum utilization of a given thyristor, are application notes and manuals published by device manufacturers. For any production design, it is necessary to work closely with the device manufacturer to be certain that the most suitable thyristors are selected and properly utilized.

REFERENCES

1. Kamahara, K., et al. 4000V 2500A High Voltage High Power Thyristor. Toronto: IEEE Conference Record of IAS Annual Meeting, 1978, pp. 1022–1028.
2. Pelly, Brian R. Power Semiconductor Devices—A Status Review. Orlando: IEEE/IAS International Semiconductor Power Converter Conference Record, 1982, pp. 1–19.
3. Locher, Ralph E. Gate-Assisted Turn-Off SCR's—New, High Frequency Inverter SCR Family. Cincinnati: IEEE Conference Record of IAS Annual Meeting, 1980, pp. 689–694.
4. Martin, I. E. A New High-Frequency SCR for Use in Inverters and Motor Controllers. Dallas: Proceedings of Powercon 8, Power Concepts, Inc. 1981, N2-6, pp. 1–10.
5. Terasawa, Y., et al. High Power Static Induction Thyristor. Washington: IEEE International Electron Devices Meeting Conferences Record, December 1979, pp. 250–253.
6. Kishi, Keiji, et al. High Power Gate Turn-Off Thyristors (GTOs) and GTO-VVVF Inverter. Palo Alto: IEEE Power Electronics Specialists' Conference Record, 1977, pp. 268–274.
7. Nagano, T., et al. Characteristics of a 3000V, 1000A Gate Turn-Off Thyristor. Philadelphia: IEEE Conference Record of IAS Annual Meeting, 1981, pp. 750–753.
8. Tada, A., and Hagino, H. A High-Voltage, High-Power, Fast Switching Gate Turn-Off Thyristor. Orlando: IEEE/IAS International Semiconductor Power Converter Conference Record, 1982, pp. 66–73.
9. Gentry, F. E., et al. *Semiconductor Controlled Rectifiers: Principles and Applications of P-N-P-N Devices.* Englewood Cliffs: Prentice-Hall, Inc., 1964, pp. 68–77.
10. Somos, I. Switching Characteristics of Silicon Power-controlled Rectifiers, I. Turn-On Action. AIEE Transactions, Part I, Communications and Electronics 80: July 1961, pp. 320–326.
11. Mackintosh, I. M. The Electrical Characteristics of Silicon P-N-P-N Triodes. Proceedings of the IRE 40: June 1958, pp. 1229–1235.
12. Aldrich, R. W., and Holonyak, N., Jr. Multiterminal P-N-P-N Switches. Proceedings of the IRE 46: June 1958, pp. 1236–1239.
13. Muss, D. R., and Goldberg, C. Switching Mechanism in the N-P-N-P Silicon Controlled Rectifier. IEEE Transactions on Electron Devices ED-10: May 1963, pp. 113–120.
14. *GE SCR Manual,* 6th ed. Syracuse: General Electric, 1979, pp. 71–91.
15. *RCA Designer's Handbook—Solid State Power Circuits.* Somerville: RCA Corporation, 1971, pp. 220–223.
16. *GE SCR Manual,* 6th ed. Op. cit., pp. 60–62.
17. Bedford, B. D., and Hoft, R. G. *Principles of Inverter Circuits.* New York: John Wiley & Sons, 1964, p. 40.
18. *GE SCR Manual,* 6th ed. Op. cit., pp. 68–69.
19. Gutzwiller, F. W., and Sylvan, T. P. Power Semiconductor Ratings under Transient and Intermittent Loads. AIEE Transactions, Part I, Communications and Electronics: January 1961, pp. 699–706.
20. *GE SCR Manual,* 6th ed. Op. cit., pp. 35–41.

PROBLEMS

1. Consider the half-wave phase-controlled rectifier of Fig. 3P-1 using a single GE C434 SCR. Assume a case temperature $T_C = 65°C$, 800-lb mounting

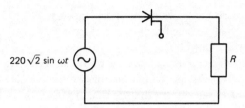

Fig. 3P-1. Half-wave-controlled rectifier.

force, double-side cooling, a 0.25-μF, 5-Ω snubber, and full-on operation such that SCR conduction occurs for each complete positive half-cycle.

a) For a power-supply frequency of 60 Hz, determine the maximum allowable peak on-state current and the corresponding watts dissipated in the SCR.

b) Calculate the RMS current and the average current through the SCR for the peak current determined in (a).

c) Use Figure 11 of the C434/C435 specification to estimate the watts dissipated in the SCR for the current found in (a).

d) Use the dc thermal resistance and the maximum allowable junction temperature from the C434/C435 specification to estimate the maximum watts dissipation permissible for the conditions in this problem.

e) Repeat (a), for 1000 Hz.

f) Repeat (a), for 10,000 Hz.

g) Select a suitable heat sink from Fig. 2P-2 for the watts dissipation determined in (f), assuming an ambient air temperature of 25°C. (Note: Two identical heat sinks will be used since double-side cooling is assumed.)

2. Consider the full-wave-controlled rectifier of Fig. 3-5 using two GE C434 SCRs. Assume $V_m = 220\sqrt{2}$, $T_c = 65°C$, 800-lb mounting force, double-side cooling, a 0.25-μF, 5-Ω snubber, and full-on operation such that each SCR conducts for a full half-cycle.

a) For a power-supply frequency of 60 Hz, determine the maximum allowable peak on-state current and the corresponding watts dissipated in each SCR.

b) Calculate the RMS current and the average current through each SCR for the peak current determined in (a).

c) Calculate the average current through the load resistor.

d) Repeat (a), for 1000 Hz and 10,000 Hz.

e) Repeat (a) and (d), for $T_c = 90°C$.

3. Use the GE C434/C435 transient thermal impedance curve (Figure 15 on specification) to determine the approximate peak half-cycle surge-current capability operating at 60 Hz and assuming an initial junction temperature

of 25°C. (Hint: Use Figure 11 of the specification to determine an "average" forward drop for an assumed peak surge current.)

4. Repeat Problem 3, for 400 Hz.

5. Use the transient thermal impedance method with repetitive pulses to check the maximum power dissipation permissible for the GE C434 SCR, as given by Figure 7 and Figure 10 on the specification sheets. Assume a 400-Hz trapezoidal-current waveform with the duty cycle $d = 0.5$, a case temperature $T_c = 90°C$, a switching voltage $V_s = 400$ V, the $di/dt = 100$ A/μs, 2000-lb mounting force, double-side cooling, and a 0.25-μF, 5-Ω snubber. (Hint: Use

$$P = \frac{T_{J,\text{MAX}} - T_C}{dR_{\theta JC} + (1 - d)Z_{\theta JC}(T + dT) - Z_{\theta JC}(T) + Z_{\theta JC}(dT)}$$

where

$$R_{\theta JC} \triangleq \text{dc thermal resistance from junction to case}$$

$$Z_{\theta JC}(T + dT) \triangleq \text{transient thermal impedance at time equal to } T + dT$$

$$Z_{\theta JC}(T) \triangleq \text{transient thermal impedance at time } T$$

$$Z_{\theta JC}(dT) \triangleq \text{transient thermal impedance at time } dT$$

$$T \triangleq \text{period}$$

$$T_{J,\text{MAX}} \triangleq \text{maximum operating junction temperature}$$

$$T_C \triangleq \text{case temperature}$$

and use Figure 15 to determine the transient thermal impedances and Figure 11 to estimate the on-state voltage.)

4
IMPORTANT CIRCUIT AND
COMPONENT CONCEPTS

4.1. INTRODUCTION

The object of this chapter is to present background information on rather elementary circuits, components, and concepts which are particularly important in power electronics. Although many subjects could be included, the four topics presented are believed to be especially significant in the development of a complete understanding of modern power electronic circuits.

4.2. *RLC* NETWORK

The series resonant circuit of Fig. 4-1 is frequently an important part of forced commutated thyristor inverters and dc-dc converters, as well as transistor series resonant converters. The differential equation for this circuit may be solved in the following straightforward manner using Laplace transforms:

$$E = L\frac{di}{dt} + \frac{1}{C}\int_0^t i\,dt + v_c(0) + Ri \qquad (4\text{-}1)$$

$$\frac{E}{s} = \left(sL + \frac{1}{sC} + R\right)I(s) - Li(0) + \frac{v_c(0)}{s} \qquad (4\text{-}2)$$

$$I(s) = \frac{\{[E - v_c(0)]/s\} + Li(0)}{sL + (1/sC) + R} = \frac{\{[E - v_c(0)]/L\} + si(0)}{[s + (R/2L)]^2 + [(1/LC) - (R^2/4L^2)]} \qquad (4\text{-}3)$$

Three cases arise for particular circuit parameters.

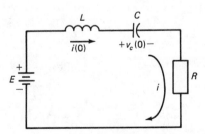

Fig. 4-1. Series resonant circuit.

Case I: Critically Damped (two identical real poles).

$$\frac{1}{LC} = \frac{R^2}{4L^2} \quad \text{or} \quad R = 2\sqrt{\frac{L}{C}} \tag{4-4}$$

$$I(s) = \frac{\{[E - v_c(0)]/L\} + si(0)}{[s + (R/2L)]^2} \tag{4-5}$$

$$i(t) = \left[\frac{E - v_c(0)}{L}\right]t\varepsilon^{(-Rt/2L)} + \left[1 - \frac{R}{2L}t\right]i(0)\varepsilon^{(-Rt/2L)} \tag{4-6}$$

$$v_c(t) = E - L\frac{di}{dt} - Ri$$

$$= E - [E - v_c(0)]\left[1 + \frac{R}{2L}t\right]\varepsilon^{(-Rt/2L)} + \frac{R^2}{4L}i(0)t\varepsilon^{\left(\frac{-Rt}{2L}\right)} \tag{4-7}$$

Case II: Over Damped (two real poles that are not equal).

$$\frac{R^2}{4L^2} > \frac{1}{LC} \quad \text{or} \quad R > 2\sqrt{\frac{L}{C}} \tag{4-8}$$

$$I(s) = \frac{\{[E - v_c(0)]/L\} + si(0)}{(s + P_1)(s + P_2)} \tag{4-9}$$

$$i(t) = \left[\frac{E - v_c(0)}{L(P_2 - P_1)}\right][\varepsilon^{-P_1 t} - \varepsilon^{-P_2 t}] + \frac{i(0)}{P_2 - P_1}[P_2\varepsilon^{-P_2 t} - P_1\varepsilon^{-P_1 t}] \tag{4-10}$$

$$v_c = E - L\frac{di}{dt} - Ri = E - \frac{[E - v_c(0)]}{P_2 - P_1}(P_2 t^{-P_1 t} - P_1\varepsilon^{-P_2 t})$$

$$- \frac{i(0)}{C(P_2 - P_1)}(\varepsilon^{-P_2 t} - \varepsilon^{-P_1 t}) \tag{4-11}$$

Case III: Under Damped or Oscillatory (complex pair of poles).

$$\frac{R^2}{4L^2} < \frac{1}{LC} \quad \text{or} \quad R < 2\sqrt{\frac{L}{C}} \tag{4-12}$$

$$i(t) = \left[\frac{E - v_c(0)}{\omega L}\right] \varepsilon^{(-Rt/2L)} \sin \omega t + i(0)\varepsilon^{(-Rt/2L)}\left(\cos \omega t - \frac{R}{2\omega L} \sin \omega t\right)$$

$$(4\text{-}13)$$

or

$$i(t) = \frac{[E_2 - v_c(0)]}{\omega L} \varepsilon^{(-Rt/2L)} \sin \omega t - \frac{\omega_o}{\omega} i(0)\varepsilon^{(-Rt/2L)} \sin(\omega t - \phi) \qquad (4\text{-}14)$$

where

$$\omega \triangleq \sqrt{\frac{1}{LC} - \frac{R^2}{4L^2}}$$

$$\omega_o \triangleq \frac{1}{\sqrt{LC}}$$

$$\phi \triangleq \tan^{-1}\frac{2\omega L}{R}$$

$$v_c(t) = E - L\frac{di}{dt} - Ri$$

$$= E - \frac{E - v_c(0)}{\omega}\left[-\frac{R}{2L}\varepsilon^{(-Rt/2L)} \sin \omega t + \omega\varepsilon^{(-Rt/2L)} \cos \omega t\right]$$

$$+ \frac{\omega_o L i(0)}{\omega}\varepsilon^{(-Rt/2L)}\left[-\frac{R}{2L} \sin(\omega t - \phi) + \omega \cos(\omega t - \phi)\right]$$

$$- R\frac{[E - V_c(0)]}{\omega L}\varepsilon^{(-Rt/2L)} \sin \omega t + \frac{\omega_o R i(0)}{\omega}\varepsilon^{(-Rt/2L)} \sin(\omega t - \phi)$$

$$= E - \omega_o\frac{[E - v_c(0)]}{\omega}\varepsilon^{(-Rt/2L)}\left(\frac{R}{2\omega_o L} \sin \omega t + \frac{\omega}{\omega_o} \cos \omega t\right)$$

$$+ \frac{\omega_o^2 L}{\omega} i(0)\varepsilon^{(-Rt/2L)}\left[\frac{R}{2\omega_o L} \sin(\omega t - \phi) + \frac{\omega}{\omega_o} \cos(\omega t - \phi)\right]$$

$$(4\text{-}15)$$

$$v_c(t) = E - \frac{\omega_o}{\omega}[E - v_c(0)]\varepsilon^{(-Rt/2L)} \sin(\omega t + \phi) + \frac{i(0)}{\omega C}\varepsilon^{(-Rt/2L)} \sin \omega t$$

$$(4\text{-}16)$$

The explicit transient response of the *RLC* circuit is given by Eqs. (4-6) and (4-7), (4-10) and (4-11), or (4-15) and (4-16). However, it is useful to study actual response curves. Figure 4-2 shows a CSMP simulation program for the circuit of Fig. 4-1, and Figs. 4-3 through 4-7 present computer-generated plots of the response curves for a range of circuit parameters. With Q_o defined as

$$Q_o \triangleq \frac{\omega_o L}{R}$$

$$(4\text{-}17)$$

FILE: KKK   OUTPUT   X   UNIVERSITY OF MISSOURI COMPUTER NETWORK

```
* FOR GRAPH
******************************************************************
INITIAL
  NOSORT
    PARAMETER E=100.0    ,L=5.0E-5  ,C=2.0E-6  ,I0=0.0  ,EC0=0.0
    PARAMETER R=(0.05,1.0,5.0,10.0,20.0)
*   PARAMETER R=(0.5,1.0)
    FIXED N1,N2
/       DIMENSION AI(103),AT(103),AV(103)
******************************************************************
*     EC   ; CAPACITOR VOLTAGE - INITIAL VOLTAGE
*     I    ; INDUCTOR CURRENT
*     X1   ; INTEGRAL OF CURRENT
*     R    ; RESISTANCE
*     C    ; CAPACITOR
*     L    ; INDUCTANCE
*
      N1=0
      N2=4
DYNAMIC
  NOSORT
      DEC=I/C
      DI=(E-EC-R*I)/L
* SAVE FOR PLOTTER*********
      N2=N2+1
      IF(N2.LE.4)GO TO 19
      N2=1
      IF(N1.LE.101)N1=N1+1
      AI(N1)=I
      AV(N1)=EC
      AT(N1)=TIME
      IF(N1.LE.101)TIMEFF=TIME
  19 CONTINUE
****** INTEGRAL**************
      I=INTGRL(I0,DI)
      EC=INTGRL(EC0,DEC)
TERMINAL
****** FOR PLOT *************
      WRITE(11)AI,AV,AT,N1
      TIMER DELT=7.50E-7,PRDEL=1.50E-6,OUTDEL=1.50E-6,FINTIM=7.50E-5
      METHOD RKSFX
*   PRTPLT EC (I)
END
    PARAMETER R=(0.05,1.0,5.0,10.0,20.0),EC0=50.0,E=0.0
END
    PARAMETER R=(0.05,1.0,5.0,10.0,20.0),EC0=100.0,E=0.0
END
    PARAMETER R=(0.05,1.0,5.0,10.0,20.0),I0=20.0,EC0=0.0,E=100.0
END
    PARAMETER R=(0.05,1.0,5.0,10.0,20.0),I0=10.0,EC0=50.0
END
STOP
  OUTPUTS    INPUTS   PARAMS  INTEGS + MEM BLKS  FORTRAN  DATA CDS
  16(500)    32(1400) 10(400)   2+  0=  2(300)   17(600)     14
```

Fig. 4-2. CSMP simulation program for circuit in Fig. 4-1.

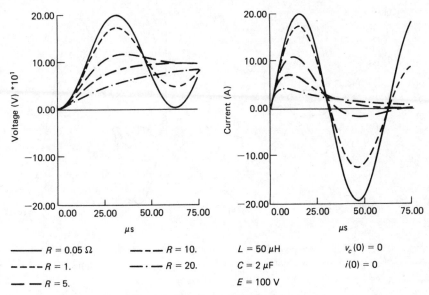

Fig. 4-3. Capacitor voltage and current.

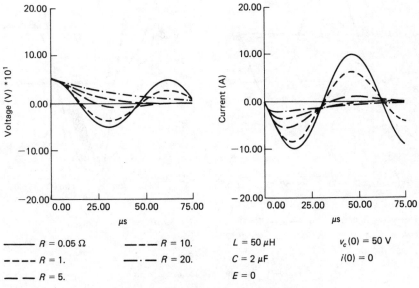

Fig. 4-4. Capacitor voltage and current.

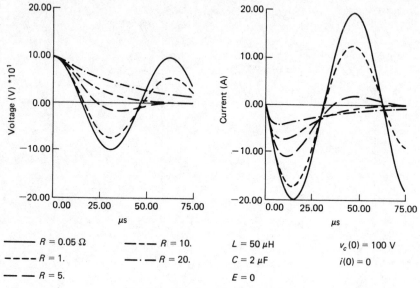

Fig. 4-5. Capacitor voltage and current.

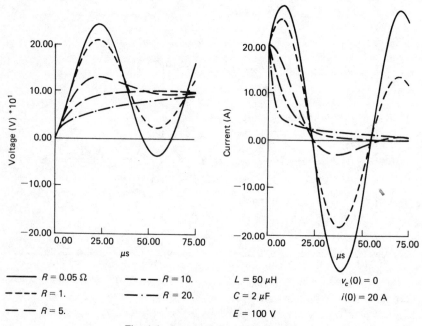

Fig. 4-6. Capacitor voltage and current.

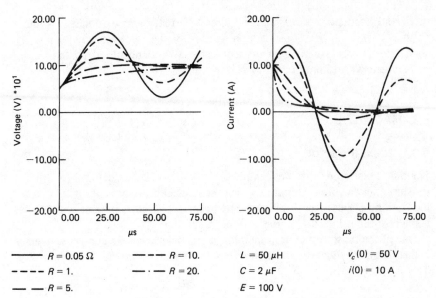

——— $R = 0.05\ \Omega$ — — — $R = 10.$ $L = 50\ \mu H$ $v_c(0) = 50$ V

— — — — $R = 1.$ — · — $R = 20.$ $C = 2\ \mu F$ $i(0) = 10$ A

— — $R = 5.$ $E = 100$ V

Fig. 4-7. Capacitor voltage and current.

the range of Q_o is from 100 to 1/4 for the resistance values from 0.5 Ω to 20 Ω. With $R = 10\ \Omega$, the circuit parameters used yield the critically damped case $(R^2/4L^2) = (1/LC)$. For this case, the damping factor δ is unity where

$$\delta \triangleq \frac{R}{2\sqrt{L/C}} = \frac{R}{2\omega_o L} = \frac{1}{2Q_o} \qquad (4\text{-}18)$$

The damping factors for other values of the resistance R are indicated on the response curves.

As a result of a study of the RLC circuit equations and response curves, several general remarks can be made:

1. The transient response is the conventional second-order-system response. It ranges from the purely oscillatory case when $R = 0$ through the damped oscillatory characteristic to the limiting situation of the simple RC circuit response as Q_o is reduced.
2. The amplitudes of the capacitor voltage and current are increased with a greater magnitude net circuit driving voltage, $[E - v_c(0)]$.
3. An initial inductor current also tends to increase the capacitor voltage and current amplitudes.

4. For all oscillatory situations, the circuit current reverses polarity within a time interval in the order of a half-period of the resonance frequency, ω_o, where the precise time of current reversal is determined by the circuit parameters and initial conditions.

4.3. MAGNETICS

4.3.1. Ideal Inductor

For this discussion, the ideal inductor is one which has negligible winding resistance and a core material in which the magnetic flux is a linear function of the winding current (see Fig. 4-8). The linear $\phi - i$ characteristic also implies a single straight-line *B-H* curve with no saturation.

One of the most important principles in practical magnetic component design is that *magnetic flux is proportional to the volt-seconds applied to the winding.* This is shown easily from

$$v = N\frac{d\phi}{dt}$$ (4-19)

Then

$$d\phi = \frac{1}{N}v\,dt$$ (4-20)

or

$$\phi(t_1) - \phi(t_0) = \frac{1}{N}\int_{t_0}^{t_1} v\,dt$$ (4-21)

More precisely, Eq. (4-21) shows that the change in flux during a time interval t_0–t_1 is proportional to the integral of the voltage over the interval, or the volt-seconds applied to the winding. Thus, if the voltage is constant over a given interval, the flux changes in a linear fashion as shown in Fig. 4-8(c) and (e). Similarly, if the applied voltage is a sine wave, the magnetic flux is a negative cosine wave since the flux is proportional to the integral of the voltage. When an ac voltage is applied to an ideal inductor, the initial flux determines the dc flux level, as illustrated in Fig. 4-8(e) and (f).

An expression for the inductance of a given coil wound on a specific core is obtained by equating the two expressions for the induced voltage in the winding.

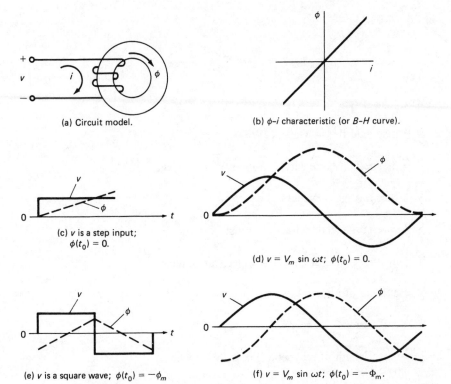

(a) Circuit model.

(b) ϕ–i characteristic (or B-H curve).

(c) v is a step input; $\phi(t_0) = 0$.

(d) $v = V_m \sin \omega t$; $\phi(t_0) = 0$.

(e) v is a square wave; $\phi(t_0) = -\phi_m$

(f) $v = V_m \sin \omega t$; $\phi(t_0) = -\Phi_m$.

Fig. 4-8. Ideal inductor. (a) Circuit model; (b) ϕ-i characteristic (or B-H curve); (c) v is a step input; $\phi(t_0) = 0$; (d) $v = V_m \sin \omega t$; $\phi(t_0) = 0$; (e) v is a square wave; $\phi(t_0) = -\phi_m$; (f) $v = V_m \sin \omega t$; $\phi(t_0) = -\Phi_m$.

$$v = L\frac{di}{dt} = N\frac{d\phi}{dt} = \frac{d\lambda}{dt} \qquad (4\text{-}22)$$

and then

$$L = N\frac{d\phi}{di} = \frac{d\lambda}{di} \qquad (4\text{-}23)$$

Using the additional relations

$$\phi = BA \qquad (4\text{-}24)$$

$$H = \frac{Ni}{l_m} \qquad (4\text{-}25)$$

in Eq. (4-23),

$$L = \frac{N^2 A}{l_m} \frac{dB}{dH} = \frac{\mu N^2 A}{l_m}$$ (4-26)

where

$\phi \triangleq$ magnetic flux — webers (Wb)

$B \triangleq$ magnetic flux density — webers/meter^2 (teslas)

$\lambda \triangleq$ flux linkage — webers

$A \triangleq$ core cross-sectional area — square meters

$H \triangleq$ magnetic field strength — ampere-turns/meter

$N \triangleq$ number of turns

$i \triangleq$ coil current — amperes

$l_m \triangleq$ mean length of magnetic flux path — meters

$\mu \triangleq$ permeability — henrys/meter ($4\pi \times 10^{-6}$ in perfect vacuum)

$L \triangleq$ inductance — henrys

An important principle in inductor design is shown by Eq. (4-26): *the inductance is directly proportional to the incremental permeability of the core material, which is the slope of the B-H curve.*

For most practical inductors, it is quite difficult to calculate the inductance precisely. However, a reasonably good approximation of the inductance can be obtained in the following cases.

Case I: Infinitely Long Air-Core Solenoid

In this case, the magnetic field strength H, as well as the flux density B, is assumed to be uniform inside the coil and in the axial direction of the coil. Immediately adjacent to the outside of the coil, H is assumed to be zero in the axial direction. Then, Ampere's law applied to the closed path indicated in Fig. 4-9(a) gives the following result:

$$\oint \mathbf{H} \cdot \mathbf{dl} = Ni$$ (4-27)

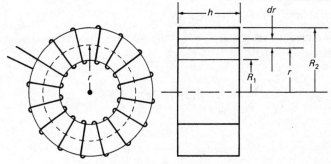

(b) Uniformly wound toroidal core.

Fig. 4-9. Inductors. (a) Long air-core solenoid; (b) uniformly wound toroidal core.

or

$$\int_0^{l_c} H\,dl + \int_{l_c}^{l_c+d} (0)\,dl + \int_{l_c+d}^{2l_c+d} (0)\,dl + \int_{2l_c+d}^{2l_c+2d} (0)\,dl = Ni \qquad (4\text{-}28)$$

and

$$Hl_c = Ni \qquad (4\text{-}29)$$

(Note: In Eq. (4-28), the second and the last integrals can be considered nearly zero, either because the components of H in the direction of the paths of length d are nearly zero, or because d is small with respect to l_c.) Then

$$L = N\frac{d\phi}{di} = N\frac{d(BA)}{di} = \mu NA\frac{dH}{di} = \frac{4\pi \times 10^{-7}N^2A}{l_c} = \frac{N^2 D_c^2 \pi^2}{l_c} \times 10^{-7}$$

$$(4\text{-}30)$$

where

$L \triangleq$ inductance in henrys

$N \triangleq$ total turns

$A \triangleq$ cross-sectional area inside of solenoid coil in square meters ($= \pi D_c^2/4$)

$D_c \triangleq$ diameter of solenoid in meters

$l_c \triangleq$ length of solenoid in meters

Case II: Uniformly Wound Toroidal Core

For this situation, shown in Fig. 4-9(b), Ampere's law is applied around the circular path with length $2\pi r$:

$$\oint \mathbf{H} \cdot \mathbf{dl} = H(2\pi r) = Ni \qquad (4\text{-}31)$$

and

$$H = \frac{Ni}{2\pi r} \qquad (4\text{-}32)$$

The total flux in the core is

$$\phi = \int_{R_1}^{R_2} \mathbf{B} \cdot \mathbf{dA} = \int_{R_1}^{R_2} Bh\,dr = \int_{R_1}^{R_2} \mu Hh\,dr = \frac{\mu Nih}{2\pi} \int_{R_1}^{R_2} \frac{dr}{r} = \frac{\mu Nih}{2\pi} \ln \frac{R_2}{R_1}$$
$$(4\text{-}33)$$

Then

$$L = N\frac{d\phi}{di} = \frac{\mu N^2 h}{2\pi} \ln \frac{R_2}{R_1} = 2 \times 10^{-7} \mu_r N^2 h \ln \frac{R_2}{R_1} \qquad (4\text{-}34)$$

where μ_r is the relative permeability of the core material.

When the magnetic path length is not constant, as is the case for the wound toroidal core, an equivalent "mean length of the magnetic path" l_m is often used to calculate the inductance. This essentially assumes that all of the magnetic flux flows around the path l_m. For the inductor of Fig. 4-9(b), L can be expressed in terms of l_m as follows:

$$L = \frac{N\phi}{i} = \frac{N(BA)}{i} = \frac{NA}{i} \mu H = \frac{\mu NA}{i} \left(\frac{Ni}{l_m}\right) = \frac{\mu N^2 A}{l_m} = \frac{\mu N^2 h(R_2 - R_1)}{l_m}$$
$$(4\text{-}35)$$

Then, l_m can be determined by equating Eqs. (4-34) and (4-35):

$$\mu \frac{N^2h}{2\pi} \ln \frac{R_2}{R_1} = \frac{\mu N^2 h (R_2 - R_1)}{l_m}$$

$$\frac{1}{2\pi} \ln \frac{R_2}{R_1} = \frac{R_2 - R_1}{2\pi R_m}$$

$$R_m = \frac{R_2 - R_1}{\ln(R_2/R_1)} \tag{4-36}$$

where R_m is the mean radius of the circular magnetic flux path. Using the series expansion for the denominator of Eq. (4-36),

$$R_m = \frac{R_2 - R_1}{[(R_2 - R_1)/R_1] + \frac{1}{2!}[(R_2 - R_1)/R_1]^2 + \frac{1}{3!}[(R_2 - R_1)/R_1]^2 + \cdots} \tag{4-37}$$

and considering only the first two terms of the denominator,

$$R_m \approx \frac{R_2 - R_1}{[(R_2 - R_1)/R_1]\{1 + \frac{1}{2}[(R_2 - R_1)/R_1]\}} = \frac{R_1}{1 + \frac{1}{2}[(R_2 - R_1)/R_1]}$$

$$\approx R_1 + \frac{1}{2}(R_2 - R_1) = \frac{R_1 + R_2}{2} \tag{4-38}$$

Thus, when $(R_2 - R_1)/R_1$ is small, the mean length of the magnetic path for the toroidal core with rectangular cross section is simply the circumference for the average radius of the core. This is a good approximation in many practical cases. Even when $(R_2 - R_1)/R_1$ is one-half, the circumference corresponding to Eq. (4-38) is less than 2 percent in error from that given by Eq. (4-36).

Case III: Gapped Core

For many dc filter inductors, an air gap is required to prevent core saturation. In this case, the magnetic circuit is the series combination of a magnetic path and an air-gap path. If the air gap is relatively small, fringing can be neglected, so that the magnetic flux density in the air gap is the same as in the magnetic core. Then Ampere's law yields

$$Hl_c + Hl_g = Ni \tag{4-39}$$

and

$$\frac{B_c}{\mu_c}l_c + \frac{B_g}{\mu_g}l_g = Ni \qquad (4\text{-}40)$$

or

$$B_c\left(\frac{l_c}{\mu_c} + \frac{l_g}{\mu_g}\right) = Ni \qquad (4\text{-}41)$$

since $B_c = B_g$. Then

$$\frac{\phi_c}{A_c}\left(\frac{l_c}{\mu_c} + \frac{l_g}{\mu_g}\right) = Ni \qquad (4\text{-}42)$$

Neglecting fringing also implies that the cross-sectional area of the core is equal to the air-gap area. Finally, from Eq. (4-42),

$$L = \frac{N\phi}{i} = N\left[\frac{NA_c}{(l_c/\mu_c) + (l_g/\mu_g)}\right] = \frac{N^2}{(l_c/\mu_c A_c) + (l_g/\mu_g A_c)} \qquad (4\text{-}43)$$

In most practical cases, the permeability of the core material is much greater than the permeability of air. However, the gap length is usually much less than the magnetic path length through the core. If the ratio μ_c/μ_g is much greater than l_c/l_g, then

$$L \approx \frac{\mu_g A_c N^2}{l_g} = \frac{4\pi \times 10^{-7} A_c N^2}{l_g} \qquad (4\text{-}44)$$

4.3.2. Transformer

The ideal transformer is illustrated in Fig. 4-10. It is assumed to have the following:

1. Negligible winding resistance.
2. Perfect coupling between windings.
3. An ideal core.

An ideal core is one for which there is negligible magnetizing current, negligible core loss, and no saturation. These assumptions imply a core with a B-H curve which is a vertical line along the $H = 0$ axis.

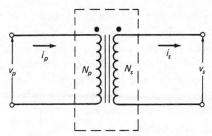

Fig. 4-10. Ideal transformer.

Two equations define the behavior of the ideal transformer.

$$\frac{v_p}{v_s} = \frac{N_p}{N_s} \tag{4-45}$$

$$N_p i_p = N_s i_s \tag{4-46}$$

These equations mean that the voltage induced in a winding is proportional to its turns or, more precisely, that the voltage induced in a secondary winding is equal to the turns ratio of the secondary to the primary times the primary voltage. Also, there is ampere-turn equality; that is, the primary ampere-turns are equal to the secondary ampere turns.

An equivalent circuit model of the practical transformer is shown in Fig. 4-11. This model is still not exact, but it is a reasonable approximation to an actual transformer in many practical situations. Figure 4-11 neglects

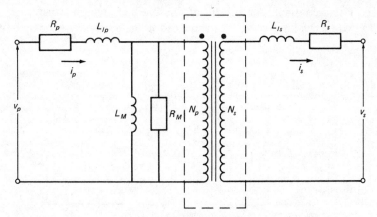

Fig. 4-11. Practical transformer.

hysteresis and saturation in the *B-H* characteristic for the magnetic core. The primary and secondary winding resistances are represented by R_p and R_s. Imperfect coupling is taken into account by the addition of lumped inductors L_{lp} and L_{ls}. These are called the leakage inductances because they account for the fact that the flux produced by current flowing in one winding does not all link the other winding. The magnetizing current is determined by the fixed inductor L_M. In actual transformers, the equivalent magnetizing inductance becomes lower when transformer saturation occurs, because this means that the slope of the *B-H* curve becomes smaller. The hysteresis and eddy current loss in the core is represented by the equivalent fixed resistance R_M. These losses are actually nonlinear functions of the applied frequency, the core flux, and the flux history, but at a given operating point, a fixed value of R_M is a reasonable approximation for the hysteresis and eddy current loss. The ideal transformer in Fig. 4-11 is required to account for the turns ratio and electrical isolation between the primary and secondary windings.

EXAMPLE

Consider the coupled magnetic circuit in Fig. 4-12. Assume zero initial conditions with the switch open. Plot the waveforms for i, i_p, i_s, and v_p, assuming that the switch is closed at $t = 0$ for 10-μs and then opened.

Figure 4-13 shows the waveforms.

Solution:

At $t = 0$ when the switch is closed, the 200-V dc source is connected across the inductor. This also applies 200 V to the primary of the ideal transformer. The transformer core flux then changes linearily in accordance with Eq. (4-22) — the derivative of the flux with respect to time must be a constant if the applied voltage v is a constant. As a result of this linear change in transformer core flux, a constant voltage of 400 V is induced in the secondary winding — posi-

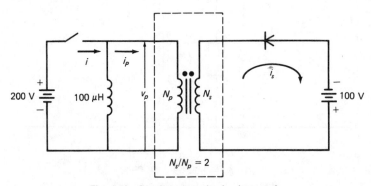

Fig. 4-12. Coupled magnetic circuit example.

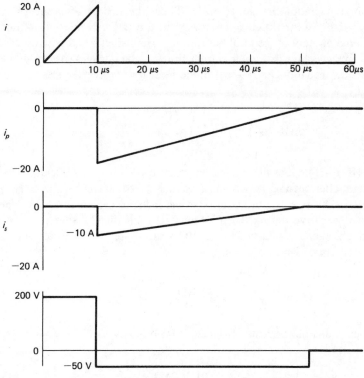

Fig. 4-13. Waveforms for circuit in Fig. 4-12.

tive at the "dot" end of the secondary. No secondary current flows because there is a reverse voltage of 500 V across the diode. Thus, i_p is zero during this interval. However, the current i increases linearly due to the constant voltage of 200 V applied to the 100 μH inductor. Again from Eq. (4-22),

$$\frac{di}{dt} = \frac{v}{L} \qquad (4\text{-}47)$$

or

$$i = \frac{1}{L} \int_0^t v\,dt + i(0) = \frac{200}{100 \times 10^{-6}}t = 2 \times 10^6 t \qquad (4\text{-}48)$$

since $i(0) = 0$.

When the switch is opened at $t = 10$ μs, the current i abruptly drops to zero since the path for this current no longer exists. However, the current through the inductor cannot change abruptly. Thus, the instant after the switch is opened, the current flowing downward in the inductor must flow upward through the primary of the ideal transformer. The inductor and primary winding form a freewheeling path with current flowing counterclockwise in this loop. Ampere-turn equality [Eq. (4-46)] in the ideal transformer requires the current $i_s(10 \ \mu s^+)$ to be

$$i_s(10 \ \mu s^+) = -\frac{N_p}{N_s} i_p(10 \ \mu s^-) = \frac{i_p(10 \ \mu s^+)}{2} \tag{4-49}$$

where $(10 \ \mu s^-)$ implies the instant before the switch is opened, and $(10 \ \mu s^+)$ is the instant after opening the switch. Note that i_p does not change from $(10 \ \mu s^-)$ to $(10 \ \mu s^+)$ because the current through the inductor cannot change instantly.

With i_s negative during the time interval immediately after the switch is opened, the diode conducts and the 100-V dc source is connected across the transformer secondary. From Eq. (4-45),

$$v_p = \frac{N_p}{N_s}(-100) = -50 \ \text{V} \tag{4-50}$$

This means that the current i_p from Eq. (4-47) is now

$$i_p = -\frac{1}{L} \int_{10 \ \mu s}^{t} v_p \, dt + i_p(10 \ \mu s^+) = \frac{(10^6 t - 10)}{2} - 20 \tag{4-51}$$

This continues until the current tries to reverse polarity, which occurs at $t = 50 \ \mu s$. At that point, the diode again blocks so that the circuit reverts to the off condition.

4.4. ASSUMED-STATE ANALYSIS

This approach is particularly useful in the analysis of rectifier and inverter circuits—to predict circuit behavior in advance of detailed analysis, as an aid in sketching possible waveforms, and to provide a means of checking some aspects of the results from computer simulations. The following two steps are required.

1. *Assume a state for the system.* This means choosing which devices are conducting and which are blocking during the particular time interval. For each device, the choice is based on the answers to the following questions. Is the device receiving a signal to turn it on? Was the device conducting previously? If the device was previously blocking, was there a forward voltage across it immediately prior to the present interval?

2. *Check that no operating principles are violated.* There must be no forward voltage across a diode which is assumed to be blocking; negative current must not be flowing through diodes or thyristors assumed to be conducting; a thyristor cannot be in the on-state unless it is being gated or it was previously conducting and still is conducting forward current; no abrupt jumps in inductor currents or capacitor voltages can occur.

The value of this method is very much a function of how well one is able to guess the state of each semiconductor device and how thoroughly one checks that no operating principles are violated. For a complex circuit involving a large number of devices, the probability of choosing the correct state for every device becomes small. Also, the time required to check all of the possible combinations becomes large, and the ability of one to check every possible violation of circuit operating principles becomes more limited. Thus, the assumed-state analysis method works best when the circuit is not too complex, and when there is a reasonably good understanding in advance of at least the general operating behavior.

The next section includes an illustration of the application of this method for a rather simple circuit. A more elaborate example is presented in Chapter 9. For polyphase ac phase-control circuits, the assumed state analysis method is quite useful.

4.5. BIASED SEMICONDUCTOR SWITCHES

This concept is based upon the superposition of currents flowing in a circuit branch. However, when the branch is an ideal undirectional semiconductor device such as a diode or reverse-blocking thyristor, there is sometimes a reluctance to apply the superposition principle.

One of the simplest examples of a biased semiconductor switch is shown in Fig. 4-14. Figure 4-15 presents curves of the currents I_1, I_2, and I_D as a function of voltage E_1. There are two possible operating states for this circuit—the diode conducting state or the diode blocking state. Using the assumed-state analysis approach of the previous section, the equations during each state are derived as follows. First, assuming the diode conducting state, there must be zero voltage across the diode and therefore the current loops for I_1 and I_2 are decoupled. Thus,

$$I_{1c} = \frac{E_1}{R_1} \tag{4-52}$$

$$I_{2c} = \frac{E_2}{R_2} \tag{4-53}$$

$$I_D = I_{1c} - I_{2c} \tag{4-54}$$

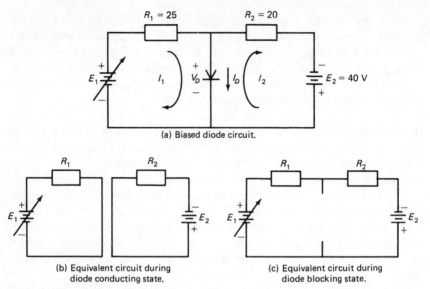

(a) Biased diode circuit.

(b) Equivalent circuit during diode conducting state.

(c) Equivalent circuit during diode blocking state.

Fig. 4-14. Biased diode (a) Biased diode circuit; (b) equivalent circuit during diode conducting state; (c) equivalent circuit during diode blocking state.

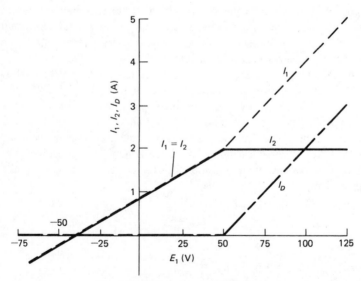

Fig. 4-15. Currents vs. E_1 for Fig. 4-14.

Equation (4-54) shows that the condition required for the diode conducting state is

$$I_{1c} > I_{2c} \tag{4-55}$$

or, from Eqs. (4-52) and (4-53),

$$\frac{E_1}{R_1} > \frac{E_2}{R_2} \quad \text{or} \quad E_1 > \frac{R_1}{R_2}E_2 \tag{4-56}$$

For the particular circuit parameters used in Fig. 4-14, Eq. (4-56) implies that I_D is greater than zero and thus the diode conducting state exists when E_1 is greater than 50 V.

For the diode blocking state, the diode branch is open so that the equivalent circuit is as shown in Fig. 4-14(c). In this case,

$$I_{1B} = I_{2B} = \frac{E_1 + E_2}{R_1 + R_2} \tag{4-57}$$

and the voltage across the diode branch V_D is given by the expression

$$V_D = E_1 - I_{1B}R_1 = E_1 - \frac{E_1 + E_2}{R_1 + R_2}R_1 \tag{4-58}$$

For the circuit parameters used,

$$V_D = E_1 - \frac{E_1 + 40}{25 + 20}(25) = \frac{4E_1 - 200}{9} \tag{4-59}$$

and this shows that V_D is negative, as required when E_1 is less than 50 V.

For simplicity, a dc circuit was used to illustrate the biased-semiconductor switch principle. Generally the variables are time-varying quantities in practical power electronic circuits, but the principle is still applicable. Problems at the end of this chapter are simple illustrations of this sort. Also, the biased switch is often a reverse-blocking thyristor instead of a diode and thyristor.

The biased-semiconductor switch principle is most useful in the analysis of the thyristor choppers and inverters discussed in Chapters 10 and 11.

REFERENCES

1. Halliday, David, and Resnick, Robert. *Physics for Students of Science and Engineering,* 2d ed. New York: John Wiley & Sons, January 1963, chaps. 36 and 38.

2. Boast, Warren B. *Vector Fields*. New York: Harper & Row, 1964, chap. 18.
3. Lee, Reuben. *Electronic Transformers and Circuits*, 4th ed. New York: John Wiley & Sons, 1964.

PROBLEMS

1. Starting from the differential equation (4-1), show in detail the derivation of Eqs. (4-10) and (4-11).

2. Assume that the switch in the circuit of Fig. 4P-1 is closed at $t = 0$ with zero initial conditions. Determine the values of L and C to provide a resonant frequency f_0 of 10 kHz and a peak current of 100 A with $E = 250$ V.

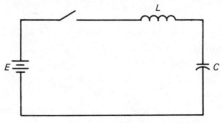

Fig. 4P-1. *L-C* circuit.

3. Consider zero initial inductor current and assume that the switch in Fig. 4P-2 is closed at $t = 0$ for 100 μs and then opened again.

a) Calculate the inductor current at $t = 100$ μs.

b) Derive the equation for the inductor current beginning the instant after the switch is opened.

c) Accurately sketch the inductor current for all time.

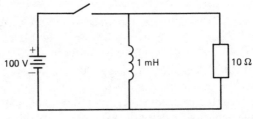

Fig. 4P-2. *L-R* circuit.

4. Consider a uniformly wound toroidal-powered iron core with a rectangular cross section where

$$l_m = 6.35 \text{ cm (mean length of magnetic path)}$$

A_c = .635 cm² (core cross-sectional area)

μ_r = 125 (relative permeability)

N = 1120 turns

a) Calculate the inductance.
b) Determine the steady-state maximum flux density in the core with a square-wave ac applied voltage which has an amplitude of 20 V and a frequency of 4000 Hz.

5. Consider a pair of tape-wound cut C-cores which are clamped together using insulating spacers in each gap, as illustrated in Fig. 4P-3. Assume that the core cross-sectional area is 6.45 cm², that the total gap length is 0.254 cm, and the $H1$ in the core is negligible.

a) Calculate the inductance with 1000 turns total wound on the cores.
b) Determine the worst-case maximum flux density in the core, assuming an RMS voltage of 110 V is applied to the winding with a frequency of 60 Hz.

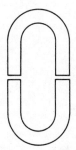

Fig. 4P-3. C-core inductor.

6. Consider the circuit of Fig. 4P-4 with an ideal transformer except assuming a ϕ-i core characteristic as in Fig. 4P-5. Assume that the switch in Fig. 4P-4

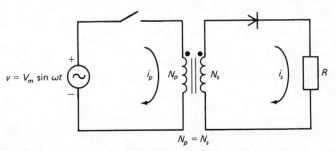

Fig. 4P-4. Transformer-rectifier circuit.

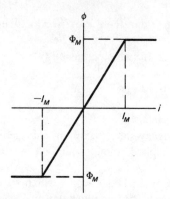

Fig. 4P-5. Modified ideal core characteristic.

is closed at the positive going zero crossing of voltage v with ϕ equal to zero and with both currents i_p and i_s zero at that instant. Carefully plot the waveforms for i_s and ϕ for enough cycles to indicate the steady state waveforms.

5
TRANSISTOR SWITCHING REGULATORS

5.1. IMPORTANCE OF SWITCHING OPERATION

As discussed in Chapter 1, possibly the most important unique aspect of power electronics is switching operation. When the transistor is operated as a switch, it is possible to control relatively large amounts of power with little transistor dissipation. In fact, with an ideal switch having negligible leakage current when off, negligible voltage drop when on, and zero switching time, there is negligible power loss in the switch. With practical transistors, the dissipation in the transistor switch can be a very small fraction of the load power controlled.

Switching transistor power supplies have become very important commercial products in the power-supply industry. They presently offer the most compact and high-performance power-regulating technique in applications with ratings up to at least several kilowatts. The maximum power is limited by the ratings of the largest bipolar and power field-effect transistors available.

The terms switching regulator, switched-mode converter, and switching power supply are all used nearly synonymously to identify these circuits. Two circuit topologies which have been widely applied are discussed in this chapter — the step-down (buck) regulator and the step-up (boost) regulator. Also, numerous additional switching regulator circuits have been developed ([1]–[4]).

5.2. STEP-DOWN (BUCK) REGULATOR

Figure 5-1 shows the ideal and a more practical version of the step-down or buck circuit. With this approach the output voltage can be controlled downward from the source voltage E by changing the on-to-off time-ratio of the switch. For the circuit of Fig. 5-1(a), the average load voltage is

$$V_{\text{LOAD, AVE}} = \frac{t_{\text{ON}}}{t_{\text{ON}} + t_{\text{OFF}}} E = \frac{t_{\text{ON}}}{T} E = dE \qquad (5\text{-}1)$$

This average load voltage may be controlled by operating at a fixed frequency with a variable on-time, operating with a constant t_{ON} using a variable switching frequency, or employing a combination of these pulse-modulation techniques to control the switching duty cycle d.

In most practical situations, the pulsed waveform of Fig. 5-1(a) is not suitable, and it is necessary to include a load filter to provide relatively low ripple dc voltage across the load. Figure 5-1(b) shows one of the simplest filter arrangements. This filter involves a linear inductance L and a coasting or free-

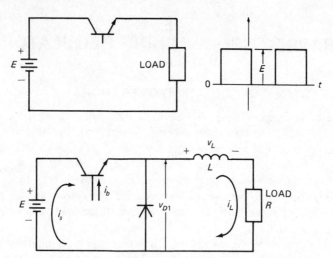

Fig. 5-1. Step-down regulator. (a) Without filter; (b) with $D1$-L filter.

wheeling diode $D1$. The waveforms for the circuit of Fig. 5-1(b), assuming a resistive load, are sketched in Fig. 5-2. During an on-interval, the load current as a function of time is obtained as follows:

$$E = Ri_L + L\frac{di_L}{dt} \tag{5-2}$$

$$\frac{E}{s} = (R + sL)I_L(s) - Li_L(0) \tag{5-3}$$

$$I_L(s) = \frac{(E/s) + Li_L(0)}{R + sL} = \frac{(E/L)}{s[s + (R/L)]} + \frac{i_L(0)}{s + (R/L)} \tag{5-4}$$

$$i_L(t) = \frac{E}{R}(1 - \varepsilon^{(-Rt/L)}) + i_L(0)\varepsilon^{(-Rt/L)} \tag{5-5}$$

When the switch is open and the load current is "coasting" through diode $D1$, the expression for the current is the same except that no driving voltage is present.

$$i_L(t) = i_L(0)\varepsilon^{(-Rt/L)} \tag{5-6}$$

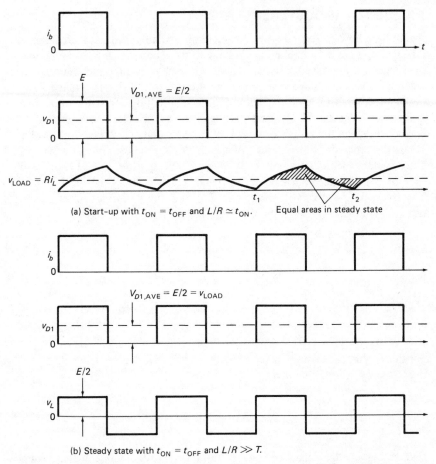

(a) Start-up with $t_{ON} = t_{OFF}$ and $L/R \simeq t_{ON}$.   Equal areas in steady state

(b) Steady state with $t_{ON} = t_{OFF}$ and $L/R \gg T$.

Fig. 5-2. Waveforms for Fig. 5-1(b). (a) Start-up with $t_{ON} = t_{OFF}$ and $L/R \simeq t_{ON}$; (b) Steady state with $t_{ON} = t_{OFF}$ and $L/R \gg T$.

For simplicity, time equal zero is defined as the beginning of the on-interval for Eq. (5-5) and then redefined as the beginning of the off-interval for Eq. (5-6).

As a result of the steady-state principle for the inductor (Section 1.5.3), there can be no average voltage across the filter inductance in steady state. Fig. 5-2(b) illustrates the waveforms, assuming steady operation with a very large filter inductance or, more specifically, $L/R \gg T$ where T is the period of the switching cycle. In this case, the large L/R time constant results in negligible ripple in the load current. The load voltage is the average value of the voltage impressed on the load L-R circuit since the average voltage across the ideal inductance must be zero in steady state.

5.3. STEP-UP (BOOST) REGULATOR

Figure 5-3 shows a second type of switching regulator. With this arrangement, it is again possible to regulate the voltage to the load by changing the on-to-off time ratio of the transistor switch. However, in this case the minimum average value of the load voltage is approximately equal to the supply voltage E. The important waveforms are sketched in Fig. 5-4, assuming negligible ripple in the inductor current and the capacitor voltage, a resistive load, and steady-state operation. When the transistor switch is closed — the on-interval — the voltage across the inductor is the supply voltage E. During the off-interval of the transistor switch, the inductor voltage is E minus the capacitor voltage. For steady-state operation, there must be zero average voltage across the ideal inductor. Therefore, the positive volt-seconds on L during the on-interval must equal its negative volt-seconds when the transistor switch is open. In general,

$$(E)(t_{ON}) = (V_{C,AVE}|_{t_{OFF}} - E)(t_{OFF}) \tag{5-7}$$

and thus

$$V_{C,AVE}|_{t_{OFF}} = \frac{t_{ON} + t_{OFF}}{t_{OFF}}E = \frac{T}{t_{OFF}}E = \frac{1}{1-d}E \tag{5-8}$$

For the special case where there is negligible ripple in the capacitor voltage, V_C is essentially constant so that

$$V_{C,AVE}|_{t_{OFF}} = V_{C,AVE} = V_C = \frac{E}{1-d} \tag{5-9}$$

That is, the average capacitor voltage during t_{OFF} is equal to the average voltage over the period T (the true steady-state average voltage), and this is equal to the essentially constant capacitor voltage.

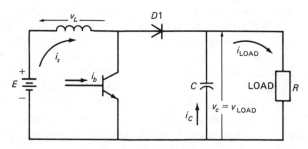

Fig. 5-3. Step-up regulator.

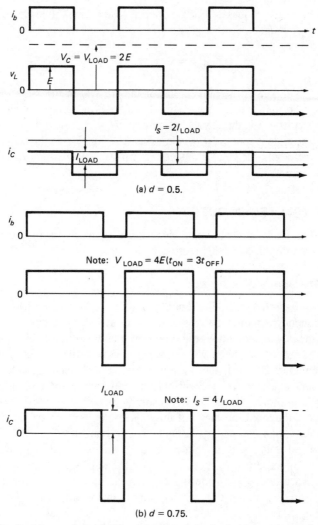

Fig. 5-4. Steady-state waveforms for Fig. 5-3 with resistive load and assuming negligible ripple in I_S and V_C. (a) $d = 0.5$; (b) $d = 0.75$.

Equations (5-8) and (5-9) clearly indicate the step-up feature of the circuit in Fig. 5-3. As the duty cycle d approaches unity, the output voltage approaches infinity. At the other extreme when the duty cycle is zero, implying that the transistor switch is closed continuously, the load voltage is equal to the source voltage. Of course, the power delivered by the source must be equal to the power consumed in the load since ideal circuit components are assumed, which means that there are no losses. For the special case of negligible ripple in I_S and V_{LOAD}, this means that

$$EI_S = V_{LOAD} I_{LOAD} \tag{5-10}$$

Thus, there is a transformer-like relationship in which the step-up ratio of voltage must be the inverse of the step-down ratio in current. A similar relationship is also true for the step-down circuit, Fig. 5-1(b), with ideal components and negligible ripple in I_{LOAD}. However, for Fig. 5-1(b) the source current i_s cannot have negligible ripple, but Eq. (5-10) still holds where I_s is the average source current.

5.4. ANALYSIS TECHNIQUES

Two important analysis techniques which are particularly useful for switching regulators are introduced in this section.

5.4.1. State-Space Averaging

This method devised by Middlebrook and Cuk [5] is very useful for analyzing the low-frequency, small-signal performance of switching circuits. It is applicable when the power switching period is short compared to the response time of the output. Thus, it has been widely applied to well-filtered, high-frequency switching circuits. The basic idea is to use state variable techniques to derive an "average" state equation which describes the input-output and control properties of the switching power converter. The resulting model is useful for studying the performance throughout the range of duty cycles, optimizing the design, determining stability, and comparing various types of converters.

The averaged state equation is developed by writing a pair of state equations for the two modes during a period T. For the subsequent discussions, the method is illustrated by considering the step-down regulator with a second-order filter, and assuming continuous current in the inductor. The circuit is shown in Fig. 5-5.

Switch Closed

$$C\frac{dv_C}{dt} + \frac{v_C}{R} = \frac{1}{L} \int (E - v_C)\, dt \tag{5-11}$$

Let

$$x_1 \triangleq \frac{1}{L} \int (E - v_C)\, dt = i_L \tag{5-12}$$

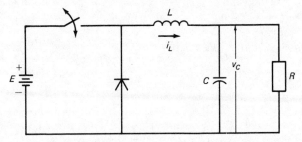

Fig. 5-5. Step-down regulator with LC filter.

$$x_2 \triangleq E - L\dot{x}_1 = v_C \qquad (5\text{-}13)$$

Then

$$\dot{x}_1 = -\frac{1}{L}x_2 + \frac{1}{L}E \qquad (5\text{-}14)$$

$$\dot{x}_2 = -L\ddot{x}_1 = \frac{dv_C}{dt} = \frac{1}{C}x_1 - \frac{1}{RC}x_2 \qquad (5\text{-}15)$$

or

$$\dot{\mathbf{x}} = \begin{bmatrix} 0 & -1/L \\ 1/C & -1/RC \end{bmatrix}\mathbf{x} + \begin{bmatrix} 1/L \\ 0 \end{bmatrix}E \qquad (5\text{-}16)$$

$$y = \begin{bmatrix} 0 & 1 \end{bmatrix}\mathbf{x} \qquad (5\text{-}17)$$

or

$$\dot{\mathbf{x}} = A_1\mathbf{x} + \mathbf{b}_1 E \qquad (5\text{-}18)$$

$$y = \mathbf{c}_1^T\mathbf{x} \qquad (5\text{-}19)$$

where

$$A_1 \triangleq \begin{bmatrix} 0 & -1/L \\ 1/C & -1/RC \end{bmatrix}; \quad \mathbf{b}_1 \triangleq \begin{bmatrix} 1/L \\ 0 \end{bmatrix}; \quad \mathbf{c}_1 = \begin{bmatrix} 0 \\ 1 \end{bmatrix} \qquad (5\text{-}20)$$

Switch Open

$$C\frac{dv_C}{dt} + \frac{v_C}{R} + \frac{1}{L}\int v_C\, dt = 0 \qquad (5\text{-}21)$$

Again, let

$$x_1 \triangleq i_L = -\frac{1}{L}\int v_C\, dt \qquad (5\text{-}22)$$

$$x_2 \triangleq v_C = -L\dot{x}_1 \qquad (5\text{-}23)$$

Then

$$\dot{x}_1 = -\frac{1}{L}x_2 \qquad (5\text{-}24)$$

$$\dot{x}_2 = \frac{1}{C}x_1 - \frac{1}{RC}x_2 \qquad (5\text{-}25)$$

or

$$\dot{\mathbf{x}} = \begin{bmatrix} 0 & -1/L \\ 1/C & -1/RC \end{bmatrix}\mathbf{x} \qquad (5\text{-}26)$$

$$y = [0 \quad 1]\mathbf{x} \qquad (5\text{-}27)$$

or

$$\dot{\mathbf{x}} = A_2\mathbf{x} + \mathbf{b}_2 E \qquad (5\text{-}28)$$

$$y = \mathbf{c}_2^T\mathbf{x} \qquad (5\text{-}29)$$

where

$$A_2 \triangleq \begin{bmatrix} 0 & -1/L \\ 1/C & -1/RC \end{bmatrix}; \qquad \mathbf{b}_2 = \mathbf{0}; \qquad \mathbf{c}_2 = \begin{bmatrix} 0 \\ 1 \end{bmatrix} \qquad (5\text{-}30)$$

In this development, the output y is defined equal to the load or capacitor voltage.

The averaged-state equation is then a weighted average of the two state equations, Eqs. (5-18), (5-19) and (5-28), (5-29), based on the duty-cycle ratio d.

$$\dot{\mathbf{x}} = A\mathbf{x} + \mathbf{b}E \qquad (5\text{-}31)$$

where

$$A \triangleq dA_1 + (1 - d)A_2; \quad \mathbf{b} \triangleq d\mathbf{b}_1 + (1 - d)\mathbf{b}_2; \quad \mathbf{c} \triangleq d\mathbf{c}_1 + (1 - d)\mathbf{c}_2 \qquad (5\text{-}32)$$

It may be shown that the approximation made by averaging is equivalent to approximating the exponential matrices $\varepsilon^{A_1 t}$ and $\varepsilon^{A_2 t}$ by their first order linear terms so that $\varepsilon^{A_1 t} + \varepsilon^{A_2 t} \approx I + (A_1 + A_2)t$ [6]. For this example,

$$dA_1 + (1 - d)A_2 = A_1 = A_2 = A \qquad (5\text{-}33)$$

$$d\mathbf{b}_1 + (1 - d)\mathbf{b}_2 = d\mathbf{b}_1 \qquad (5\text{-}34)$$

$$d\mathbf{c}_1 + (1 - d)\mathbf{c}_2 = \mathbf{c}_1 = \mathbf{c}_2 = \mathbf{c} \qquad (5\text{-}35)$$

Thus, Eq. (5-31) can be written

$$\dot{\mathbf{x}} = \begin{bmatrix} 0 & -1/L \\ 1/C & -1/RC \end{bmatrix} \mathbf{x} + d \begin{bmatrix} 1/L \\ 0 \end{bmatrix} E \qquad (5\text{-}36)$$

$$y = [0 \quad 1]\mathbf{x} \qquad (5\text{-}37)$$

Now, it is assumed that d is perturbed a small amount from a constant value; that is,

$$d \triangleq D + \Delta d \qquad (5\text{-}38)$$

This perturbation Δd from the constant duty ratio D produces the following perturbations in $\mathbf{x}$, $\dot{\mathbf{x}}$, and y

$$\mathbf{x} = \mathbf{x}_D + \Delta\mathbf{x} \qquad (5\text{-}39)$$

$$\dot{\mathbf{x}} = \dot{\mathbf{x}}_D + \Delta\dot{\mathbf{x}} \qquad (5\text{-}40)$$

$$y = y_D + \Delta y \qquad (5\text{-}41)$$

Then

$$\dot{\mathbf{x}}_D + \Delta\dot{\mathbf{x}} = \begin{bmatrix} 0 & -1/L \\ 1/C & -1/RC \end{bmatrix} (\mathbf{x}_D + \Delta\mathbf{x}) + (D + \Delta d) \begin{bmatrix} 1/L \\ 0 \end{bmatrix} E \qquad (5\text{-}42)$$

$$y_D + \Delta y = \mathbf{c}^T(\mathbf{x}_D + \Delta\mathbf{x}) \qquad (5\text{-}43)$$

but

$$\dot{\mathbf{x}}_D = \begin{bmatrix} 0 & -1/L \\ 1/C & -1/RC \end{bmatrix} \mathbf{x}_D + D \begin{bmatrix} 1/L \\ 0 \end{bmatrix} E \qquad (5\text{-}44)$$

$$y_D = \mathbf{c}^T \mathbf{x}_D \qquad (5\text{-}45)$$

which is the unperturbed average state representation, and thus

$$\Delta\dot{\mathbf{x}} = \begin{bmatrix} 0 & -1/L \\ 1/C & -1/RC \end{bmatrix} \Delta\mathbf{x} + \Delta d \begin{bmatrix} 1/L \\ 0 \end{bmatrix} E \qquad (5\text{-}46)$$

$$\Delta y = \mathbf{c}^T \Delta\mathbf{x} \qquad (5\text{-}47)$$

Equations (5-46) and (5-47) describe the small-signal linearized behavior of the step-down regulator in Fig. 5-5. For example, the transfer function, useful for stability analysis when closed-loop voltage control of the load voltage is required, is

$$\frac{\Delta y(s)}{\Delta d(s)} = \mathbf{c}^T[(sI - A)^{-1}\mathbf{b}E] = \begin{bmatrix} 0 & 1 \end{bmatrix} \begin{bmatrix} s & 1/L \\ -1/C & s + 1/RC \end{bmatrix}^{-1} \begin{bmatrix} E/L \\ 0 \end{bmatrix}$$

$$= \begin{bmatrix} 0 & 1 \end{bmatrix} \frac{\begin{bmatrix} s + (1/RC) & 1/C \\ -1/L & s \end{bmatrix}^T \begin{bmatrix} E/L \\ 0 \end{bmatrix}}{s^2 + (1/RC)s + (1/LC)}$$

$$= \begin{bmatrix} 0 & 1 \end{bmatrix} \frac{\begin{bmatrix} s + (1/RC) & -1/L \\ 1/C & s \end{bmatrix} \begin{bmatrix} E/L \\ 0 \end{bmatrix}}{s^2 + (1/RC)s + (1/LC)}$$

$$= \frac{\begin{bmatrix} (1/C) & s \end{bmatrix} \begin{bmatrix} E/L \\ 0 \end{bmatrix}}{s^2 + (1/RC)s + (1/LC)} = \frac{E/LC}{s^2 + (1/RC)s + (1/LC)}$$

$$(5\text{-}48)$$

In this case, the transfer function for the averaged model is the same as the transfer function of the *LCR* output network with *E* constant. However, the dc gain is a function of *E* as shown by Eq. (5-48).

5.4.2. Analysis Using Difference Equations

Power electronic circuits involve discrete processes since the switching devices are opened and closed at discrete instants of time. While differential equations are used to describe continuous systems, difference equations form the basis for the mathematical models of discrete systems. Just as Laplace transforms provide a simple means of solving linear differential equations, *z*-transforms are the analogous tool for the solution of linear difference equations. This is the most rigorous approach for the general solution of switching-type circuits. However, it does become quite involved even for relatively simple circuits. A serious study of references [7] and [8] is recommended for one interested in this method of analysis for power electronic switching circuits.

REFERENCES

1. Middlebrook, R. D., and Cuk, S. *Advances in Switched-Mode Power Conversion, Vol. I.* Pasadena, CA: TESLACO, 1981.
2. Cuk, S., and Middlebrook, R. D. *Advances in Switched-Mode Power Conversion, Vol. II.* Pasadena, CA: TESLACO, 1981.
3. Wood, Peter. *Switching Power Converters.* New York: Van Nostrand Reinhold Company, 1981, chap. 3.
4. Severns, Rudy. The Design of Switchmode Converters above 100 kHz. Cupertino, CA: Intersil Application Bulletin A034, 1980.
5. Middlebrook, R. D., and Cuk, Slobodan. A General Unified Approach to Modelling Switching-Converter Power Stages. IEEE Power Electronics Specialists' Conference Record, 1976, pp. 18–34.
6. Ibid., Appendix A.
7. Casteel, Jordan. *Power Electronic Circuit Analysis Using Difference Equations.* Columbia, MO: University of Missouri–Columbia, PhD Dissertation, 1978.
8. Brown, Arthur R., and Middlebrook, R. D. Sampled-Data Modeling of Switching Regulators. IEEE Power Electronics Specialists' Conference Record, 1981, pp. 349–369.
9. *The Power Transistor and Its Environment.* Malakoff, France: Thompson-CSF Semiconductor Division, 1978, chaps. X and XI.

PROBLEMS

1. Consider the circuit of Fig. 5P-1 with such a large capacitor that there is negligible ripple in the load voltage, and assuming that $t_{ON} = t_{OFF} = 400$ μs.
    a) Carefully plot the steady-state waveforms for i_S, i_L, and i_C.
    b) Calculate the power delivered by the dc source in steady state.
2. Consider the step-down circuit of Fig. 5P-2 with $L = 100$ μH, $R = 2$ Ω, $E_1 = 250$ V, $E_2 = 50$ V, and a switching period $T = 100$ μs.
    a) Determine the duty cycle d for steady-state operation.

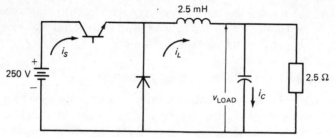

Fig. 5P-1. Step-down regulator.

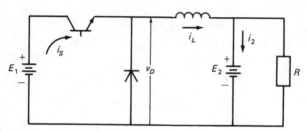

Fig. 5P-2. Step-down regulator battery charger.

b) Carefully plot the steady-state waveforms for i_S, i_L, and i_2, assuming that $i_L = 0$ at $t = 0$ when the transistor is first turned on with the duty cycle of (a).

c) Calculate the power delivered by the voltage source E_1.

d) With steady-state operation as given by (b), assume that the on-time is changed to 40 μs ($d = 0.4$) for one period T. Carefully plot the waveforms for i_S, i_L, and i_2 during the one period in which t_{ON} is 40 μs.

e) At the end of the one period during which $d = 0.4$, assume that the duty cycle is restored to the value of (a). Carefully plot the steady-state waveforms for i_S, i_L, and i_2 thereafter.

3. Consider the step-down regulator in Fig. 5-1(b) with $E = 250$ V, $R = 2.5$ Ω, and a duty cycle $d = 0.5$.

a) Calculate the steady-state average load current.

b) Assuming that $L/R = T/2$, calculate the power delivered by the dc source in steady state.

4. Consider the circuit of Fig. 5-3 with $E = 100$ V, $R = 50$ Ω, $t_{ON} = 80$ μs, and $t_{OFF} = 20$ μs. Assume that the inductance is so large that there is negligible ripple in i_S, and that C is so large that there is negligible ripple in the load voltage. For steady-state operation,

a) Calculate the load voltage.

    b) Calculate the power delivered by the 100-V source.

    c) Carefully plot the waveforms for v_L and i_C.

5. Consider the circuit of Fig. 5-3 with $E = 125$ V, $L = 125$ μH, $R = 50$ Ω, $d = 0.75$, and $T = 40$ μs. Assume that C is so large that there is negligible ripple in the load voltage. For steady-state operation,

    a) Calculate the load voltage.

    b) Calculate the power delivered by the 125-V source.

    c) Carefully plot the waveforms for v_L, i_S, and i_C.

6. Use the method of state variable averaging to determine the transfer function $[\Delta v_C(s)]/[\Delta d(s)]$ for the step-up regulator of Fig. 5-3, assuming continuous source current i_S, but otherwise arbitrary values for E, L, C, and R.

6
TRANSISTOR INVERTERS

6.1. PRINCIPLES

A rectifier circuit is used to convert ac to dc, and an inverter accomplishes the reverse process. For inverter operation, the power switching devices must have the ability to hold off forward voltage, and the interval of time when conduction occurs must be controllable [1]. The simplest inverters use switching devices which can interrupt the flow of current. Presently, such inverters utilize the bipolar junction transistor (BJT), the field-effect transistor (FET), or the gate-turn-off thyristor (GTO). SCR thyristors are used in the more complicated forced commutated inverters discussed in Chapter 11.

6.2. SINGLE PHASE

The most straightforward means of converting dc power to ac is the transistor inverter. A basic power circuit for such an inverter is shown in Fig. 6-1. These inverters have been in use for essentially as long as transistors have been available ([2]–[7]). The transistor driving circuits, not shown, alternately close one transistor switch and then the other. Thus, the dc source voltage E is alternately applied to one-half of the transformer primary and then the other to produce a square wave of voltage across the load. With ideal circuit components and with a resistive load, Fig. 6-1 is an efficient means of producing ac from a dc power source. However, components are not ideal, and loads are almost never purely

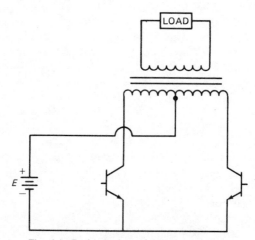

Fig. 6-1. Basic transistor inverter power circuit.

110

resistive. With a nonideal transformer and with inductive loads, the circuit of Fig. 6-1 must be modified. A simple modification to permit reliable operation with inductive loads is shown in Fig. 6-2.

The important steady-state waveforms for Fig. 6-2 are shown in Fig. 6-3, assuming a purely inductive load but still an ideal transformer and semiconductors (see Problem 1). With the addition of diodes $D1$ and $D2$, this transistor inverter can handle all types of inductive load. However, it is still necessary to add additional circuit elements to tolerate leakage reactances in the primary windings of the transformer and/or inductance in the leads to the transistors. A common approach is to connect a series $R\text{-}C$ or a more complicated snubber network across each transistor. This network is selected to minimize the transient voltages appearing across the transistors during each switching interval.

With regard to the transistor base drive circuits, it should be noted that carefully designed drive circuits are required to minimize switching losses without excessive saturated voltage drop. Numerous analog, digital, and hybrid approaches are used, and an increasing number of monolithic IC components are being produced, which include on a single chip many of the base-drive and inverter control functions required. The base-drive and control circuits are a crucial part of the design of reliable transistor inverters.

The two additional most important single-phase arrangements are shown in Fig. 6-4 and Fig. 6-5. For simplicity, each switching device is designated by the simple switch symbol in the remainder of this chapter. Each switch has a diode connected in parallel with it to provide paths for inductive load currents during switching.

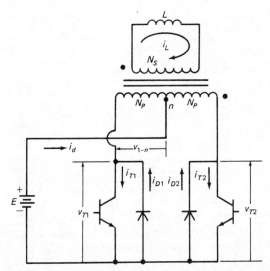

Fig. 6-2. Modified transistor inverter with L-load.

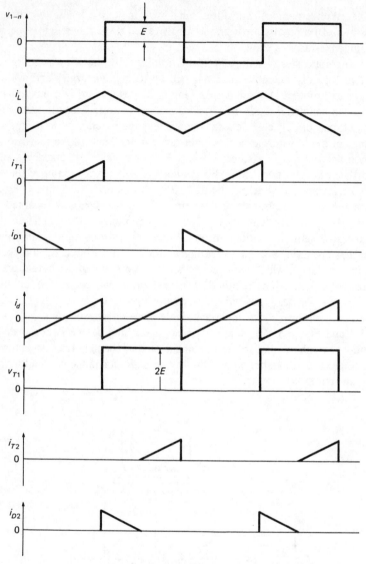

Fig. 6-3. Steady-state waveforms for Fig. 6-2 ($N_P = N_S$).

For the subsequent discussions in this section, the circuit of Fig. 6-5 is considered. It is possibly the most widely used single-phase arrangement, and the output-voltage waveforms in Figs. 6-2 and 6-4 are the same. With the simplest control, Fig. 6-5 delivers a square-wave voltage to the load. Switches $S1$ and

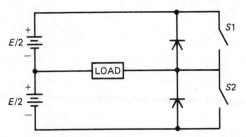

Fig. 6-4. Single-phase half-bridge inverter.

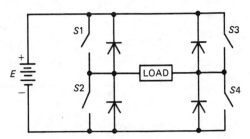

Fig. 6-5. Single-phase bridge inverter.

$S4$ are closed for one half-cycle to deliver one polarity of voltage to the load. Then these two switches are opened and $S2$ and $S3$ are closed to connect E across the load in the opposite direction. The output frequency is set by the inverter switch operating frequency. The amplitude of the square-wave load voltage is E. This is the simplest inverter control scheme, but it has a significant disadvantage. A sinusoidal output voltage should be delivered by the ideal inverter. The square-wave output has appreciable harmonic content, which is undesirable for many loads. For example, harmonics produce additional losses in a motor supplied from an inverter, and they do not supply significant useful torque. In fact, certain harmonics result in some negative torque.

A Fourier analysis of the square wave is carried out in Chapter 1. With $t = 0$ chosen as the beginning of a positive half-cycle, this square wave is an odd function with half-wave symmetry. Thus, the Fourier series must contain only odd harmonic sine terms:

$$v_L = \frac{4E}{\pi} \left[\sin \omega t + \frac{1}{3} \sin 3\omega t + \frac{1}{5} \sin 5\omega t + \frac{1}{7} \sin 7\omega t + \cdots \right] \quad (6\text{-}1)$$

A spectral presentation of Eq. (6-1) is shown in Fig. 6-6.

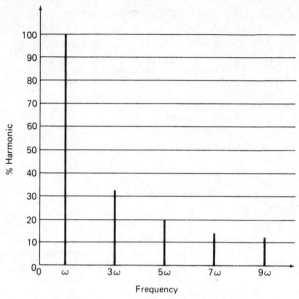

Fig. 6-6. Harmonic spectrum of square wave.

6.3. THREE PHASE

The two basic three-phase inverter arrangements are shown in Figs. 6-7 and 6-8. There are many additional polyphase circuits, but they can be derived from the arrangements in these figures. Again, the diodes in parallel with

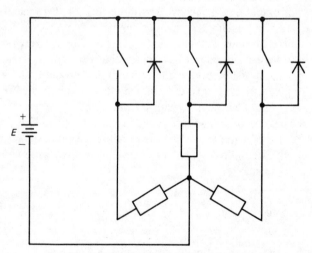

Fig. 6-7. Three-phase half-wave inverter.

each switch are necessary to provide paths for inductive load current during switching.

The three-phase bridge inverter is the most widely used. It has two operating modes — 180° conduction in each switch or 120° conduction. The 180° conduction mode is generally preferred for two reasons: 1) the utilization of the semiconductor switches is better with 180° conduction, and 2) the output-voltage waveform is effected by the load with 120° conduction.

Waveforms for the circuit of Fig. 6-8 are shown in Fig. 6-9, assuming the following:

1. 180° conduction in each switch — this implies 180°-long base drive signals for transistor switches.
2. Balanced resistive load.
3. The zero voltage reference is the midpoint of E.

For the resistive-load situation, the diodes do not conduct and thus are not required. However, with inductive loads, the diodes are essential.

It is interesting to note the change in the waveforms in Fig. 6-9 when the ac neutral point n is connected to the midpoint "0" of E. Voltages v_{a-0}, v_{b-0}, v_{c-0},

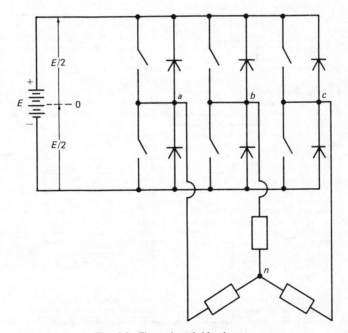

Fig. 6-8. Three-phase bridge inverter.

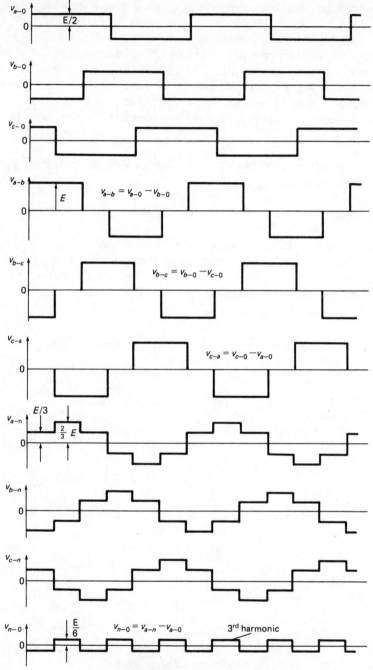

Fig. 6-9. Three-phase bridge-inverter waveforms.

v_{a-b}, v_{b-c}, and v_{c-a} are the same. However,

$$v_{a-n} = v_{a-0}$$

$$v_{b-n} = v_{b-0}$$

$$v_{c-n} = v_{c-0}$$

$$v_{n-0} = 0 \tag{6-2}$$

and a third harmonic square wave of current flows in the $n - 0$ connection. The amplitude of this current is $(E)/(2R)$, where R is the resistance in each load phase.

The harmonic spectra of the line-to-line and line-to-neutral voltage waveforms shown in Fig. 6-9 are the same as for the square wave (Fig. 6-6), except that no third harmonics or multiples of third harmonics are present (see Problems 3 and 4).

6.4. OUTPUT VOLTAGE CONTROL AND HARMONIC REDUCTION

6.4.1. dc Input Voltage Control

Control of the dc input voltage is the most straightforward means of controlling the output voltage from an inverter. For a given switching pattern, the output voltage is directly proportional to the inverter input voltage. If the power source is ac, a simple phase-controlled rectifier can be used. Whereas if dc power is available, it is necessary to use a switching-converter voltage control.

The two main advantages of dc input voltage control are simplicity and that the harmonic content of the inverter output waveform is fixed. However, a disadvantage is that the fixed harmonic content of a square wave is large, and it is thus quite expensive to filter the output waveform when low total harmonic distortion is required. Another disadvantage is that all of the power must be controlled by another converter — the dc voltage control.

6.4.2. Phase-Shift Voltage Control

Voltage control by phase shift involves phasor addition of the outputs of two or more inverters. The resulting output voltage can be controlled from zero, when the two inverter outputs exactly cancel, to a voltage equal to the sum of the inverter outputs when they are added in-phase. The simplest form of this type of voltage control may be explained by reference to the bridge inverter of Fig. 6-5. However, consider this circuit as two half-bridge inverters whose outputs are subtracted. Specifically, let v_{a-0} be the voltage from the left side of the load to a center tap on the dc input. The input is not a center-tapped source, but it is useful to consider the zero-voltage reference point as the midpoint of the dc input. The voltage v_{b-0} is the voltage from the right end of the load to the midpoint of the dc source. Thus, the inverter output voltage is

$$v_L = v_{a-0} - v_{b-0} \tag{6-3}$$

Figure 6-10 shows the waveforms for Fig. 6-5 considered as two half-bridge inverters with various phase angles between them. Actually, the inverter output

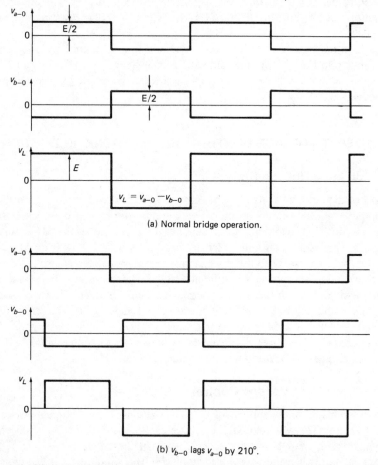

(a) Normal bridge operation.

(b) v_{b-0} lags v_{a-0} by 210°.

Fig. 6-10. Single-phase bridge-inverter output for various phase angles between left and right half-bridges. (a) Normal bridge operation; (b) v_{b-0} lags v_{a-0} by 210°; (c) v_{b-0} lags v_{a-0} by 240°; (d) v_{b-0} lags v_{a-0} by 270°; (e) v_{b-0} lags v_{a-0} by 300°; (f) v_{b-0} lags v_{a-0} by 330°; (g) v_{b-0} lags v_{a-0} by 360°.

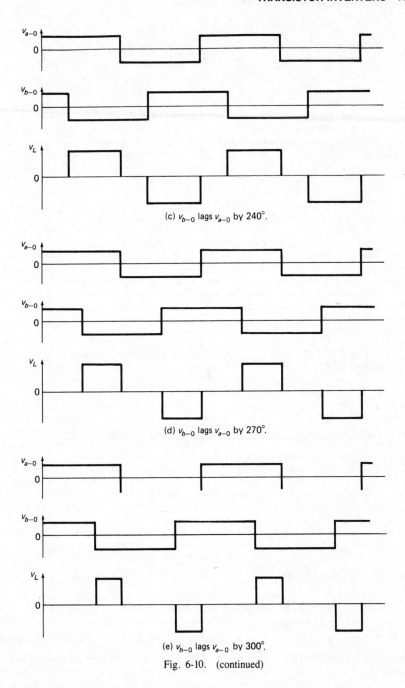

(c) v_{b-0} lags v_{a-0} by 240°.

(d) v_{b-0} lags v_{a-0} by 270°.

(e) v_{b-0} lags v_{a-0} by 300°.

Fig. 6-10. (continued)

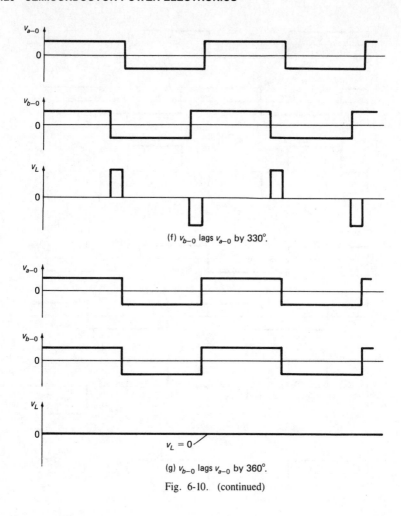

(f) v_{b-0} lags v_{a-0} by 330°.

(g) v_{b-0} lags v_{a-0} by 360°.

Fig. 6-10. (continued)

voltage is a single-phase pulse-width modulated waveform, which is discussed in Section 6.4.3.

Of course, this voltage-control technique can be extended to three phase. An obvious additional extension is to add the outputs of m-inverters. This is done either by phasor addition in a single-output transformer with m-primary windings or by connecting the secondary windings of m-separate transformers in series [8].

6.4.3. Pulse-width Modulation (PWM) Voltage Control

A somewhat more complex but very advantageous means of inverter output-voltage control is pulse-width modulation. Two classes of PWM schemes are possible: single-pulse modulation and multiple-pulse sinusoidal modulation.

Figure 6-11 is a plot of the fundamental and first three harmonic components of the single-pulse PWM wave expressed as a fraction of the fundamental component in a square wave. Figure 6-12 is a similar plot except that the harmonics are expressed as a percentage of the fundamental. The single-pulse PWM scheme provides an effective means of controlling the fundamental inverter output voltage [9]. This is a very good method of voltage control for a reasonable output range. However, when the output voltage is reduced to a low value, the harmonics become large compared with the fundamental component. Also, at reduced output voltage, the conduction angle for the switch is quite short, which generally results in poorer utilization of semiconductor switches.

Sinusoidal-pulse modulation is the most sophisticated means of controlling inverter output voltage. A technique for practically implementing such a scheme was first described by Schonung and Stemmler [10]. Numerous additional extensions of this technique have been made. Figure 6-13 illustrates the approach described in [11]. It involves a fixed-amplitude high-frequency trian-

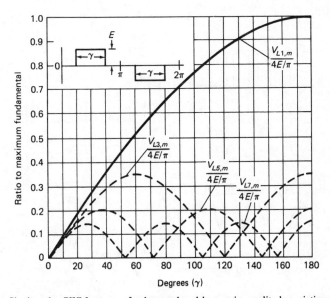

Fig. 6-11. Single-pulse PWM wave—fundamental and harmonic amplitude variations with pulse width.

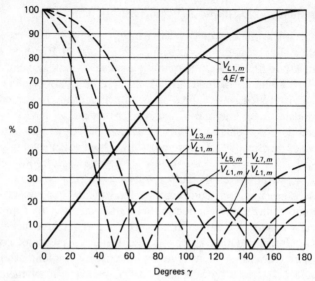

Fig. 6-12. Single-pulse PWM wave harmonic content as percentage of fundamental.

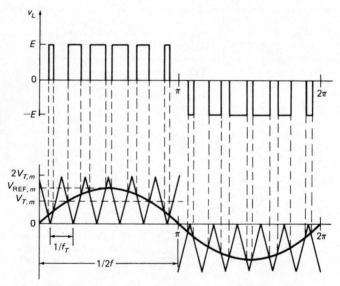

Fig. 6-13. Sinusoidal pulse modulation ($N = f_T/2f = 6$).

gular wave, which is reversed in polarity on each half-cycle and compared to an adjustable amplitude sine-wave reference serving as the input command to establish the inverter output voltage. For this type of PWM, all harmonics of order $n < 2N$ are eliminated, assuming that $V_{REF,m}/V_{T,m} < 2$. As the amplitude of the reference sine wave is changed, the fundamental component of the inverter output is varied.

6.4.4. Sinusoidal Reference-Feedback Voltage Control

Figure 6-14 is a functional diagram of this method of voltage control ([12]–[14]). It is a type of sinusoidal PWM, where the switching period is automatically varied throughout each output cycle to control the error between a sinusoidal voltage reference and the actual load voltage. The amplitude of the load voltage is established by the reference amplitude. This scheme functions to control the magnitude of the output voltage and to force the instantaneous output to follow a sinusoidal command. A PWM waveform similar to that in Fig. 6-13 appears as the input voltage to the load filter in Fig. 6-14. The variation in switching rate during the output cycle is determined by the design of the hysteresis element and the characteristics of the output filter and load. This technique is relatively insensitive to load-induced harmonics resulting from nonlinear types of loads.

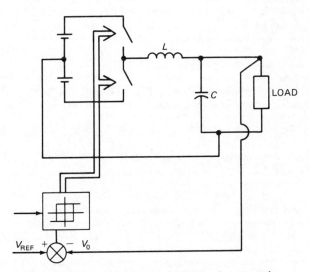

Fig. 6-14. Sinusoidal reference-feedback voltage control.

6.4.5. Harmonic Reduction

As indicated previously, the harmonic content in the basic single-phase inverter output waveform — the square wave — is rather large. Thus, in many applications it is necessary either to filter the output, which may be expensive, or to use other means to reduce the output-voltage harmonics. All of the techniques described in Sections 6.2.2–6.2.4 also may be used to improve output waveforms. In addition, the harmonic elimination approach originated by Turnbull [15] and generalized by Patel ([16] and [17]) can be quite useful. In this approach, the switching instants in a PWM waveform are set to the value which eliminate selected harmonics. Modern microprocessors with large memory capability make harmonic elimination schemes more practical.

REFERENCES

1. Bedford, B. D., and Hoft, R. G. *Principles of Inverter Circuits,* New York: John Wiley & Sons, 1964.
2. Roddam, Thomas. *Transistor Inverters and Converters.* Princeton, NJ: D. Van Nostrand, 1963.
3. *Westinghouse Silicon Power Transistor Handbook.* Youngwood, PA: Westinghouse Electric, 1967, pp. 5–1 through 5–25.
4. *Semiconductor Power Circuits Handbook.* Phoenix, AZ: Motorola Semiconductor Products, 1968, chap. II.
5. *RCA Designer's Handbook — Solid State Power Circuits.* Somerville, NJ: RCA Corporation, 1971, pp. 300–338.
6. Norris, Bryan, ed. *Power Transistor and TTL Integrated-Circuit Applications.* New York: McGraw-Hill Book Company, 1977, chap. X.
7. Severns, Rudy, and Armijos, Jack, eds. Santa Clara, CA: Siliconix, 1984, pp. 6–71 through 6–80, 6–139 through 6–144.
8. Bedford, B. D., and Hoft, R. G. Op. cit., pp. 263–278.
9. Ibid., pp. 235–254.
10. Schonung, A., and Stemmler, H. Static Frequency Changers with "Subharmonic" Control in Conjunction with Reversible Variable Speed AC Drives. The Brown Boveri Review: August/September 1964, pp. 555–577.
11. Dewan, S. B., and Straughen, A. *Power Semiconductor Circuits.* New York: Wiley & Sons, 1975, pp. 406–409.
12. Kawamura, A. and Hoft, R. Instantaneous Feedback Controlled PWM Inverter with Adaptive Hysteresis. IEEE Transactions on Industry Applications, vol. IA-20, no. 4, July/August 1984, pp. 769–775.
13. Kernick, A., Stechschulte, D. L., and Shireman, D. W. Static Inverter with Synchronous Waveform Synthesized by Time-Optimal Response Feedback. IEEE Transactions on Industrial Electronics and Control Instrumentation, vol. IECI-24, 1977, pp. 297–305.
14. Fink, Donald G., and Beaty, H. Wayne, eds. *Standard Handbook for Electrical Engineers,* 11th ed. New York: McGraw-Hill, 1978, pp. 13-40 and 13-41.
15. Turnbull, F. G. Selected Harmonic Reduction in Static DC-AC Inverters. IEEE Transactions on Communications and Electronics, vol. 83, July 1964, pp. 374–378.

16. Patel, H. S., and Hoft, R. G. Generalized Techniques of Harmonic Elimination and Voltage Control in Thyristor Inverters: Part I — Harmonic Elimination. IEEE Transactions on Industry Applications, vol. IA-9, no. 3, May/June 1973, pp. 310–317.
17. Patel, H. S., and Hoft, R. G. Generalized Techniques of Harmonic Elimination and Voltage Control in Thyristor Inverters: Part II — Voltage Control Techniques. IEEE Transactions on Industry Applications, vol. IA-10, no. 5, September/October 1974, pp. 666–673.

PROBLEMS

1. Assume that the circuit in Fig. 6-2 is first turned on at the beginning of the half-cycle of conduction for transistor $T1$ and with zero initial load current i_L. Plot the steady-state waveforms corresponding to those in Fig. 6-3. Could these waveforms be observed in a practical circuit?

2. Consider the circuit of Fig. 6-2 with diodes $D1$ an $D2$ removed, but with a resistor R connected in parallel with the load inductance L. Assume that $E = 100$ V, $R = 10\ \Omega$, $L = 2.5$ mH, and an inverter operating frequency of 1000 Hz. Carefully sketch the steady-state waveforms for v_{1-n}, i_d, i_o (the total current in the transformer secondary), i_R (the current through the load resistance), and i_L (the current through the load inductance). Note: Since the load R and L are in parallel, $i_o = i_R + i_L$.

3. Determine the Fourier series representation for the line-to-line load voltage in Fig. 6-9 and draw the harmonic spectrum similar to Fig. 6-6.

4. Repeat Problem 3 for the line-to-neutral load voltage in Fig. 6-9.

5. Use the digital computer to determine the harmonic content of v_L in Fig. 6-13, assuming that $V_{\text{REF},m} = 2V_{T,m}$.

7
PHASE-CONTROLLED RECTIFIERS AND
LINE-COMMUTATED INVERTERS

7.1. REVIEW OF DIODE RECTIFIERS

As an introduction to a discussion of phase-controlled rectifiers, it is instructive to review the operation of simple diode circuits. One of the most basic circuits is shown in Fig. 7-1. This is referred to as a full-wave rectifier because an output voltage is supplied on both half-cycles of the ac input. The term *single way* implies that current flows in only one direction in the transformer secondary windings. Figure 7-1 is also called a two-pulse rectifier since there are two dc output current pulses per cycle of the ac input frequency. Another name for the circuit of Fig. 7-1 is the single-phase, center-tap rectifier.

The output voltage v_o of the circuit in Fig. 7-1 is the well-known full-wave rectified waveform shown in Fig. 7-2, assuming ideal diodes and an ideal transformer. When the dc circuit inductance is negligible, the current i_d is also the same waveform. The addition of the inductance L_d causes filtering of the load current, and this waveform then becomes as indicated in Fig. 7-2. Figure 7-3 shows a modification of Fig. 7-1 to include the effect of transformer and/or ac source reactance, also referred to as commutating reactance.* Without inductors L_{C1} and L_{C2}, it is possible for the current to transfer from one diode to

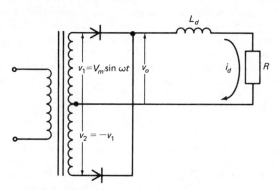

Fig. 7-1. Single-phase, full-wave, single-way rectifier.

---

*The commutation interval is the time, or angle, required to transfer conduction from one rectifying element to the next. For rectifier circuits, the commutating angle is also referred to as the "overlap angle," since it is the interval when there is overlap of conduction, i.e., the interval when both devices conduct in Fig. 7-3.

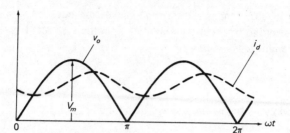

Fig. 7-2. Waveforms for Fig. 7-1.

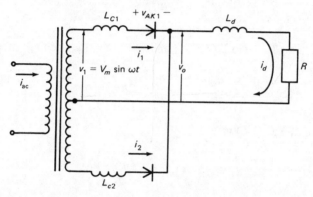

Fig. 7-3. Single-phase, center-tap rectifier with commutating-inductance.

the next, essentially instantaneously, at the zero crossings of the ac voltage v_1. When the commutating inductors are present, there cannot be abrupt jumps in the diode currents. A general expression for the commutating angle μ can be derived as follows. Considering $\omega t = 0$ at the positive-going zero crossing of v_1 in Fig. 7-3 and steady-state operation, commutation starts μ_s degrees after $\omega t = 0$. Assuming an initial dc current of $i_d(\mu_s)$, the following equations hold during the commutation interval when the current i_2 is reduced to zero and i_1 is increased from zero to the dc current value.

$\underline{\mu_s < \omega t < \mu_f}$

$$v_1 = L_C \frac{di_1}{dt} + L_d\left(\frac{di_1}{dt} + \frac{di_2}{dt}\right) + R(i_1 + i_2) \qquad (7\text{-}1)$$

$$-v_1 = L_C \frac{di_2}{dt} + L_d\left(\frac{di_1}{dt} + \frac{di_2}{dt}\right) + R(i_1 + i_2) \qquad (7\text{-}2)$$

It is assumed that the commutating inductances are equal ($L_{C1} = L_{C2} = L_C$). Angle μ_f is the point at which commutation is complete, which means that the commutating angle μ is given by

$$\mu = \mu_f - \mu_s \tag{7-3}$$

The two equations below are always true for Fig. 7-3.

$$i_1 + i_2 = i_d \tag{7-4}$$

$$v_o = L_d\left(\frac{di_1}{dt} + \frac{di_2}{dt}\right) + R(i_1 + i_2) = L_d\frac{di_d}{dt} + Ri_d \tag{7-5}$$

Substituting Eq. (7-4) into Eqs. (7-1) and (7-2), and then adding, gives

$$L_C\frac{di_d}{dt} + 2L_d\frac{di_d}{dt} + 2Ri_d = 0 \tag{7-6}$$

The solution of Eq. (7-6) is

$$i_d = i_d\left(\frac{\mu_s}{\omega}\right)\varepsilon^{[-2R/(L_C+2L_d)](t-\mu_s/\omega)} \tag{7-7}$$

Using Eq. (7-7) in Eq. (7-5) yields

$$v_o = L_d\left[\left(\frac{-2R}{L_C + 2L_d}\right)i_d\left(\frac{\mu_s}{\omega}\right)\varepsilon^{[-2R/(L_C+2L_d)](t-\mu_s/\omega)}\right] + Ri_d\left(\frac{\mu_s}{\omega}\right)\varepsilon^{[-2R/(L_C+2L_d)](t-\mu_s/\omega)}$$

$$= \frac{RL_Ci_d(\mu_s/\omega)}{L_C + 2L_d}\varepsilon^{[-2R/(L_C+2L_d)](t-\mu_s/\omega)} \tag{7-8}$$

Whenever the upper diode in Fig. 7-3 is conducting,

$$v_1 - v_o = L_C\frac{di_1}{dt} \tag{7-9}$$

The solution for $i_1(\theta)$, where θ is an arbitrary angle during the commutation period ($\mu_s < \theta < \mu_f$), is obtained by substitution of Eq. (7-8) into Eq. (7-9) and integrating as follows:

$$\frac{di_1}{dt} = \frac{v_1 - v_o}{L_C} \tag{7-10}$$

$$i_1(\theta) = \int_{\mu_s}^{\theta} \frac{v_1 - v_o}{L_C} \, d(\omega t)$$

$$= \frac{1}{L_C} \int_{\mu}^{\theta} \left[V_m \sin \omega t - \frac{R L_C i_d(\mu_s)}{L_C + 2L_d} \varepsilon^{[-2R(\omega t - \mu_s)/\omega(L_C + 2L_d)]} \right] d(\omega t)$$

$$= \frac{V_m}{\omega L_C} (\cos \mu_s - \cos \theta) + \frac{1}{2} i_d(\mu_s) \left[\varepsilon^{[-2R(\theta - \mu_s)/\omega(L_C + 2L_d)]} - 1 \right] \quad (7\text{-}11)$$

At the conclusion of this commutation, the current i_1 becomes equal to the dc current i_d, so that

$$i_1(\mu_f) = i_d(\mu_f) \quad (7\text{-}12)$$

and from Eq. (7-4) this means that

$$i_2(\mu_f) = 0 \quad (7\text{-}13)$$

Replacing θ by μ_f in Eq. (7-11) and setting the resulting expression equal to Eq. (7-7) with $t = \mu_f/\omega$ gives

$$\frac{V_m}{\omega L_C} (\cos \mu_s - \cos \mu_f) + \frac{i_d(\mu_s)}{2} \left\{ \varepsilon^{[-2R(\mu_f - \mu_s)/\omega(L_C + 2L_d)]} - 1 \right\}$$

$$= i_d(\mu_s) \varepsilon^{[-2R(\mu_f - \mu_s)/\omega(L_C + 2L_d)]} \quad (7\text{-}14)$$

or

$$\frac{V_m}{\omega L_C} (\cos \mu_s - \cos \mu_f) = \frac{i_d(\mu_s)}{2} \left\{ 1 + \varepsilon^{[-2R(\mu_f - \mu_s)/\omega(L_C + 2L_d)]} \right\} \quad (7\text{-}15)$$

Unfortunately, it is not possible to write an expression for μ_f from Eq. (7-15) in terms of the circuit parameters. It would be straightforward to determine μ_f, either by using an iterative computer solution, or simply by plotting each side of Eq. (7-15) as a function of μ to find the value that satisfies this relation, assuming that μ_s were known. However, μ_s must be determined by finding the point at which v_1 becomes greater than v_o to start the flow of current i_1.

When i_1 is zero, the voltage v_{AK1} across the upper diode of Fig. 7-3 is

$$v_{AK1} = 2v_1 + L_C \frac{di_2}{dt} = 2v_1 + L_C \frac{di_d}{dt} \quad (7\text{-}16)$$

Thus, current i_1 starts to flow when

$$2v_1 > -L_C \frac{di_d}{dt} \qquad (7\text{-}17)$$

Equations (7-2), (7-4), and (7-5), with i_1 equal to zero, yield

$$-v_1 = L_C \frac{di_d}{dt} + v_o \qquad (7\text{-}18)$$

or

$$-L_C \frac{di_d}{dt} = v_1 + v_o \qquad (7\text{-}19)$$

Thus, Eq. (7-17) can also be written

$$2v_1 > v_1 + v_o \qquad v_1 > v_o \qquad (7\text{-}20)$$

Equation (7-20) implies that v_o must be positive after $\omega t = 0$ in order for the beginning of commutation to be delayed until after the zero crossing of v_1.

In general, with continuous dc current, v_o also must be positive during commutation. This is shown as follows. Using Eq. (7-5) in Eq. (7-6) yields

$$L_C \frac{di_d}{dt} + 2v_o = 0 \qquad (7\text{-}21)$$

From Eq. (7-6),

$$\frac{di_d}{dt} = -\frac{2RL_d}{L_C + 2L_d} \qquad (7\text{-}22)$$

From Eq. (7-21),

$$2v_o = -L_C \frac{di_d}{dt} \qquad (7\text{-}23)$$

Equation (7-22) shows that di_d/dt is negative with nonzero dc current. Thus, Eqs. (7-22) and (7-23) together imply that v_o is positive during commutation. This is one of the differences between the circuit of Fig. 7-3 and the single-phase bridge circuit. In the latter case, v_o is zero during commutation since all four diodes conduct, which essentially shorts the v_o output terminals during commutation.

Digital simulation is the most practical way to determine μ_f, μ_s, and μ for arbitrary circuit parameters. However, Eq. (7-15) can be simplified when the dc circuit inductance L_d is large enough such that there is negligible variation in the dc current during the commutation interval. This implies that the dc current is essentially constant during commutation, and thus Eq. (7-4) becomes

$$i_1 + i_2 = I_{d\mu} \text{ (a constant)} \tag{7-24}$$

and then differentiating,

$$\frac{di_1}{dt} + \frac{di_2}{dt} = 0 \quad \text{or} \quad \frac{di_1}{dt} = -\frac{di_2}{dt} \tag{7-25}$$

During commutation when both diodes are conducting in Fig. 7-3,

$$2v_1 = L_C \frac{di_1}{dt} - L_C \frac{di_2}{dt} \tag{7-26}$$

Combining Eqs. (7-25) and (7-26) results in

$$2v_1 = 2L_C \frac{di_1}{dt} \quad \text{or} \quad v_1 = L_C \frac{di_1}{dt} = -L_C \frac{di_2}{dt} \tag{7-27}$$

Substituting Eq. (7-27) into Eq. (7-9) shows that $v_o = 0$ during the commutation interval. This also means that μ_s is zero from the comment following Eq. (7-20). In other words, commutation starts at the zero crossing of v_1, i.e., at $\omega t = 0$, and μ_f is equal to μ. Thus, Eq. (7-15) becomes

$$\frac{V_m}{\omega L_C}(1 - \cos \mu) = I_{d\mu} \tag{7-28}$$

The exponential term in Eq. (7-15) is approximately unity with the assumption of negligible variation in the dc current during the commutation interval. Thus, the right side of Eq. (7-15) reduces to $i_d(0) = I_{d\mu}$. From Eq. (7-28), the expression for the commutation angle becomes

$$\cos \mu = 1 - \frac{\omega L_C I_d}{V_m} \tag{7-29}$$

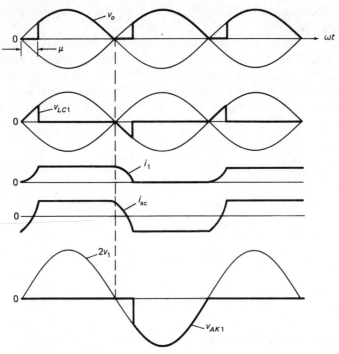

Fig. 7-4. Waveforms for Fig. 7-3 assuming negligible ripple in the dc current.

The important steady-state waveforms are shown in Fig. 7-4 assuming negligible change in the dc current during commutation. In fact, the dc current is assumed to have negligible ripple, which means it is essentially constant throughout each cycle for the waveforms in Fig. 7-4.

7.2. SINGLE-PHASE-CONTROLLED RECTIFIER

7.2.1. Resistive Load

The circuit in Fig. 7-5 is the single-phase, center-tap-controlled rectifier. The rectifying elements are reverse-blocking triode thyristors (SCRs) instead of diodes, as was the case for the circuits discussed in Section 7.1. Figure 7-6 illustrates the important waveforms for phase-control angles of $\alpha = 60°$ and $\alpha = 120°$.*

---

*The phase control angle α is the delay angle or the angle by which conduction is retarded, measured from the angle which produces the maximum rectifier dc voltage. For the single-phase-controlled rectifier of Fig. 7-5, $\alpha = 0°$ means that SCR gating occurs at the ac voltage-zero crossings, so that conduction begins at the same instants that it would if diode rectifying elements were used.

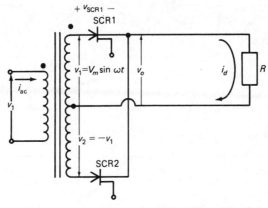

Fig. 7-5. Single-phase, center-tap-controlled rectifier with resistive load.

7.2.2. *L-R* Load

Figure 7-7 is the same circuit as in Fig. 7-5 except that an *L-R* load is used.

With the application of phase-control gating signals to the SCRs, it is possible to control the average load voltage and current. Essentially all rectifier circuits can be used as phase-controlled rectifiers if SCR devices are employed. Figure 7-8 illustrates the important steady-state waveforms for the circuit of Fig. 7-7, assuming discontinuous current operation at a gating angle of $\alpha = 60°$.* Throughout this text "continuous" gating generally will be used rather than "pulse" gating. This means that the gate drive circuits are assumed to provide a gating voltage during the entire period for possible conduction of a particular thyristor. Short gating pulses of the order of 10 μs duration are adequate in many simple phase-control applications. However, at least multiple pulses are sometimes essential. Thus, to preserve the greatest generality, continuous gating is used throughout this text.

Figure 7-9 shows typical steady-state waveforms for the circuit of Fig. 7-7, assuming the continuous-current mode of operation at a gating angle of $\alpha = 60°$. It should be noted that the shaded areas in Fig. 7-8 or Fig. 7-9 must be equal to satisfy the steady-state principle for the inductor.

Two additional angle definitions must be mentioned for phase-controlled rectifiers. While α is the angle when conduction begins during a particular operating period, the symbol β is used to designate the angle when conduction stops through a rectifying element. For the continuous-current case shown in Fig. 7-9, the angle $\beta = \alpha + \pi$. However, for the discontinuous-current case of Fig. 7-8, $\beta < \alpha + \pi$. The conduction angle is designated γ, and $\gamma = \beta - \alpha$.

---

*The term *discontinuous current* means that the periodic dc current is zero for a finite interval of time during each period. The dc current may be zero for an instant during each period, or it may never be zero, when the dc current is continuous.

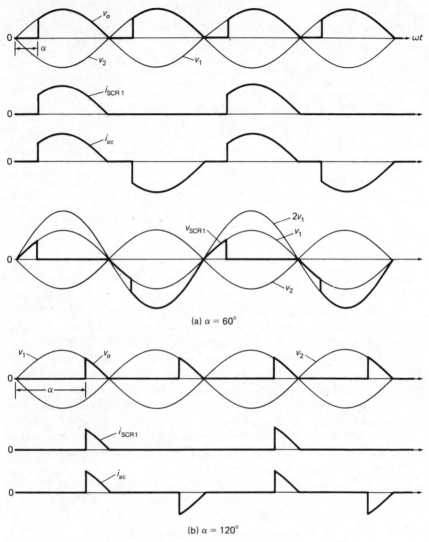

(a) $\alpha = 60°$

(b) $\alpha = 120°$

Fig. 7-6. Waveforms for Fig. 7-5. (a) $\alpha = 60°$; (b) $\alpha = 120°$.

7.2.3. CEMF Load

Figure 7-10 shows the single-phase, center-tap-controlled rectifier used as a dc motor drive. It is assumed that a separately circuited dc field is used. The series *L-R* in the armature may just represent the armature circuit impedance. However, a dc filter reactor is also often used to smooth the dc motor current.

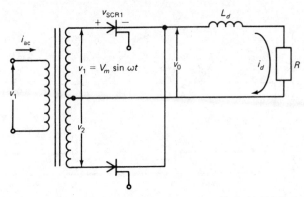

Fig. 7-7. Single phase, center-tap-controlled rectifier with R-L load.

Figure 7-11 shows steady-state waveforms, assuming discontinuous current, and Fig. 7-12 assumes continuous current. In either case, it is assumed that there is negligible speed change during a cycle so that the motor counter-EMF is essentially constant.

7.2.4. Effects of ac Line Reactance

The effects of commutating or ac circuit reactance on phase-controlled rectifiers are similar to the effects on diode rectifiers discussed in Section 7.1, assuming continuous dc current. However, commutation begins when the next SCR is gated rather than at the zero crossings of the ac voltage waveform. Thus, the commutating angle μ is a function of α, since it is a function of the ac voltage at the time of commutation. If the dc circuit current i_d is assumed essentially constant during the commutation interval, a similar expression to Eq. (7-24) may be developed as follows. During the particular commutation when SCR1 is gated, after SCR2 has been conducting,

$$\frac{di_{\text{SCR1}}}{dt} = \frac{v_1}{L_C} \tag{7-30}$$

$$i_1(\alpha + \mu) = i_d(\alpha) = I_{d\mu} = \frac{1}{\omega L_C} \int_{\alpha}^{\alpha+\mu} V_m \sin \omega t \, d(\omega t)$$

$$= \frac{V_m}{\omega L_C} [-\cos \omega t] \Big|_{\alpha}^{\alpha+\mu} = \frac{V_m}{\omega L_C} [\cos \alpha - \cos(\alpha + \mu)] \tag{7-31}$$

$$\cos(\alpha + \mu) = \cos \alpha - \frac{\omega L_C I_{d\mu}}{V_m} \tag{7-32}$$

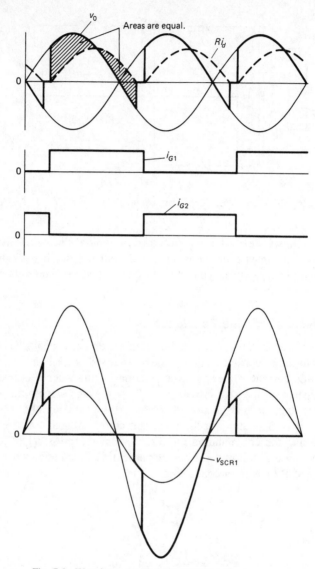

Fig. 7-8. Waveforms for Fig. 7-7 with discontinuous current.

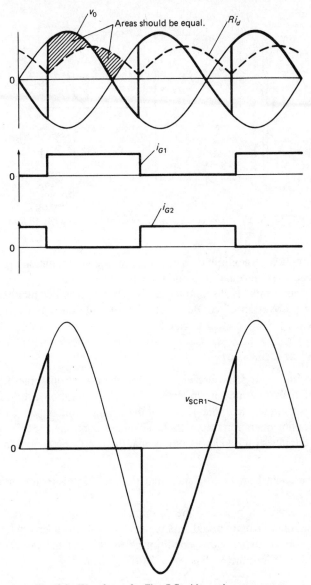

Fig. 7-9. Waveforms for Fig. 7-7 with continuous current.

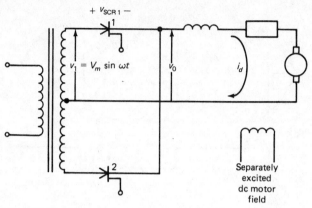

Fig. 7-10. Single-phase, center-tap-controlled rectifier with CEMF load.

For the assumption of negligible change in i_d during commutation, again the dc output voltage v_o is zero during commutation.

When the dc current is discontinuous, commutation is not required ($\mu = 0$ and there is no overlap of conduction), since the dc current is already zero when the next SCR to conduct is gated.

7.2.5. Inverter Operation

Another very interesting and useful feature of phase-controlled rectifiers is that they may also be operated as inverters. Inverter operation implies that the power flow is from the dc to the ac side of the circuit — the opposite of the situation for rectifier operation. Essentially all phase-controlled rectifiers may be operated in the inverting mode, assuming that the following conditions are met:

1. The phase control angle α is such that the rectifier average output voltage is negative.
2. There is a source of power in the dc circuit.
3. The ac voltage is maintained independent of the inverter operation. (An alternate way to express this condition is to say that the inverter must be connected to an infinite ac bus, or a "stiff" ac bus.)

As a means of introducing the concepts of inverter operation, consider the simple case of two dc voltages, as shown in Fig. 7-13. In Fig. 7-13(a), it is assumed that $E_1 > E_2$, which means that the current I flows in the clockwise direction. This results in a flow of power from dc voltage E_1 to E_2. For Fig. 7-13(b), the voltage E_2 is assumed to be greater than E_1, which produces the opposite direction of power flow. If E_1 and E_2 are considered ideal bat-

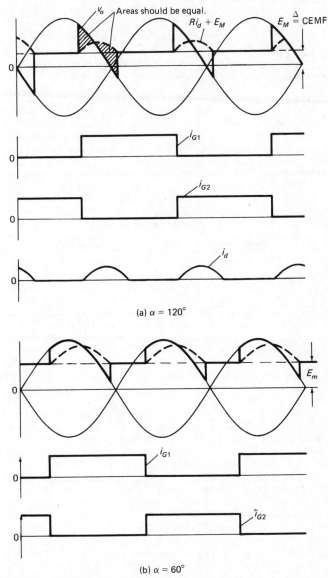

Fig. 7-11. Waveforms for Fig. 7-10 with discontinuous current. (a) $\alpha = 120°$; (b) $\alpha = 60°$.

teries, then it is clear that a given battery is supplying power (being discharged) when current is flowing out of its positive terminal. On the other hand, when current is flowing into the positive terminal of a particular battery, it is receiving power (being charged).

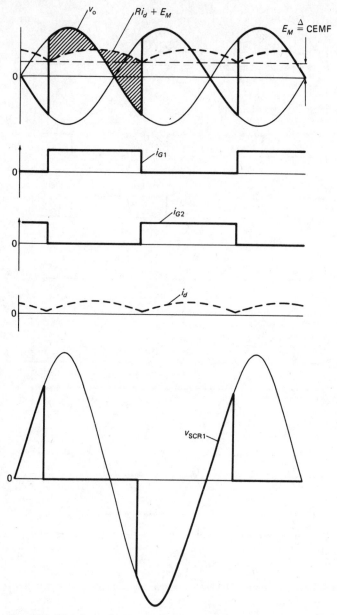

Fig. 7-12. Waveforms for Fig. 7-10 with continuous current and $\alpha = 60°$.

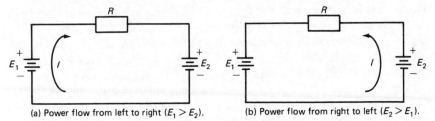

(a) Power flow from left to right $(E_1 > E_2)$.   (b) Power flow from right to left $(E_2 > E_1)$.

Fig. 7-13. dc power flow. (a) Power flow from left to right $(E_1 > E_2)$; (b) power flow from right to left $(E_2 > E_1)$.

The next simplest case is shown in Fig. 7-14. Two sinusoidal voltage sources are connected in series opposition, with a resistor R also included in the series-connected network. If the voltage v_1 has a larger amplitude than v_2, a net current i flows, where

$$i = \frac{v_1 - v_2}{R} \tag{7-33}$$

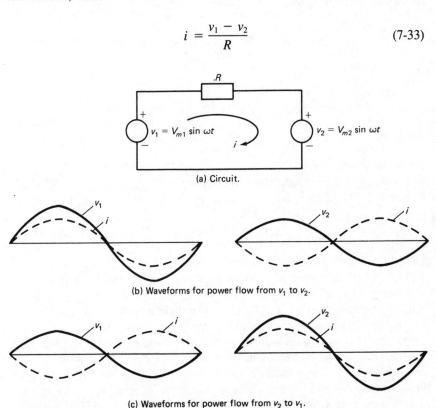

(a) Circuit.

(b) Waveforms for power flow from v_1 to v_2.

(c) Waveforms for power flow from v_2 to v_1.

Fig. 7-14. ac power flow. (a) Circuit; (b) waveforms for power flow from v_1 to v_2; (c) waveforms for power flow from v_2 to v_1.

This current is in phase with the voltage v_1, as shown in Fig. 7-14(b), and 180° out of phase with voltage v_2. Following the same convention used with dc sources, an ac voltage source is delivering power when the current flows out of the source terminal which is instantaneously positive. On alternate half-cycles, the polarities of both the voltage and current, for a particular source, reverse, but the power flow is still in the same direction. For example, on the negative half-cycle of v_1, the lower terminal of this source in Fig. 7-14 becomes positive, but the current also reverses so that current is again flowing out of the positive terminal of source v_1 on the negative half-cycle. Figure 7-14(c) illustrates the situation opposite that shown in Fig. 7-14(b). The amplitude of the sinusoidal voltage of source v_2 is greater than that of v_1, so the power flow is from source v_2 to v_1.

Figure 7-15 shows a somewhat more general case where a series R-L is included in the network. If voltage source v_1 has a greater amplitude than v_2, a net current will flow, given by the following:

$$i = \frac{V_{m1} - V_{m2}}{\sqrt{R^2 + (\omega L)^2}} \sin(\omega t - \theta) \tag{7-34}$$

where

$$\theta = \tan^{-1} \frac{\omega L}{R} \tag{7-35}$$

In this case, the current which flows lags the voltage v_1 by the angle θ, since the circuit impedance is inductive. The power flows from source v_1 to v_2, where the power delivered by source v_1 is given by

$$P_1 = V_{1,e} I_e \cos \theta \tag{7-36}$$

It should be noted that the current could be considered leading voltage v_2 by an angle of $\pi - \theta$. However, with simple R, L, and C impedances, the power factor of the current delivered by a source will be either leading or lagging by an angle of less than or equal to 90°. Thus, the angle $\pi - \theta$ will be greater than 90°, and this indicates that source v_2 is receiving power.

The final situation which needs to be mentioned is the phase-controlled rectifier with a large dc circuit inductance. In this case, the current flowing from the ac source approaches a square wave, as shown in Fig. 7-16. The fundamental component of the ac source current will lag the ac source voltage v. However, it is also possible to deliver power to the ac side of a phase-controlled rectifier. This leads to the simplest example of inverter operation of phase-controlled rectifiers.

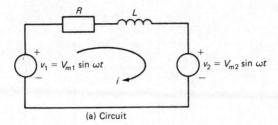

(a) Circuit

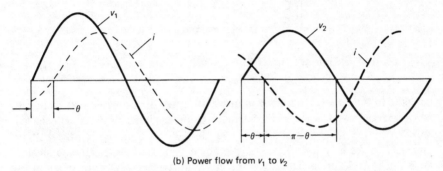

(b) Power flow from v_1 to v_2

Fig. 7-15. ac power flow with L-R impedance. (a) Circuit; (b) power flow from v_1 to v_2.

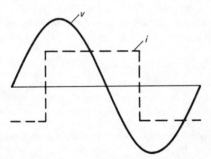

Fig. 7-16. ac voltage and current waveforms with large dc inductance for single-phase-controlled rectifier.

Consider the circuit of Fig. 7-17. For simplicity, it is assumed that the dc circuit inductance L_d is so large that there is negligible ripple in the current i_d. Also, the resistance in the dc circuit is assumed negligible, and the turns ratio of the transformer is considered 1:1 from the primary to each half of the secondary. These assumptions are not essential to the operation of the phase-controlled rectifier/inverter, but they simplify the waveforms. The resulting steady-state waveforms are shown in Fig. 7-18. The average value of v_o must

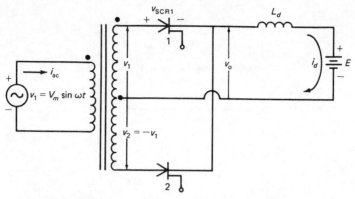

Fig. 7-17. Single-phase, center-tap-controlled rectifier battery charger.

be equal to the battery voltage E in order to satisfy the principle that there can be no average voltage across the inductor L_d in steady state. Thus, the shaded areas in Fig. 7-18 above and below the dotted line representing the battery voltage E must be equal. It should be noted that when $\alpha = 90°$, the shaded areas in Fig. 7-18 are equal only if the dc circuit voltage E is zero. In addition, if α is greater than $90°$, then the dc circuit voltage must be negative to satisfy the steady-state principle of the inductor. Figure 7-19 illustrates the waveforms for an α of $150°$.

Assuming continuous dc circuit current, but not necessarily negligible ripple, the expression for the average value of v_o as a function of firing angle α is as follows:

$$V_o = \frac{1}{\pi} \int_\alpha^{\pi+\alpha} V_m \sin \omega t \, d(\omega t) = \frac{2V_m}{\pi} \cos \alpha \qquad (7\text{-}37)$$

This expression is plotted in Fig. 7-20. If the situation here is compared with the two dc sources of Fig. 7-13, the regions of rectifier and inverter operation are clear. Consider the phase controlled rectifier average output voltage as an equivalent dc source. When this source is positive, power flows from it to the load battery. This is rectifier operation, and it occurs for $0 < \alpha < 90°$. When the equivalent dc voltage source, representing V_o, is negative, the battery voltage must also be negative. This is inverter operation, which occurs for $90° < \alpha < 180°$. It is most important to note that the power flow reversed because of a reversal in the polarity of the two dc voltages, V_o and E, while the direction of the dc current flow remained the same.

The shape of the curve in Fig. 7-20 is the same for all single-phase- and polyphase-controlled rectifiers considered in this text, assuming that the dc cur-

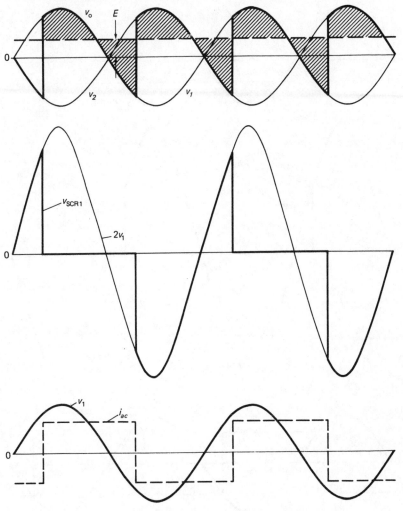

Fig. 7-18. Waveforms for Fig. 7-17 ($\alpha = 60°$).

rent is continuous. The maximum rectifying and inverting voltage magnitude is different depending upon the particular current. However, the load circuit may be essentially any combination of linear and nonlinear impedances and voltage sources, provided that the dc output current from the phase-controlled rectifier inverter is continuous. Inverter operation is also possible when the dc circuit current is not continuous. However, then the curve corresponding to that in Fig. 7-20 is different for particular circuits and specific operating conditions.

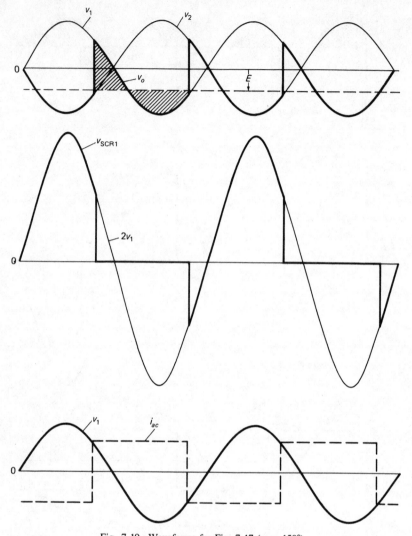

Fig. 7-19. Waveforms for Fig. 7-17 ($\alpha = 150°$).

Two types of power factors are defined for situations when nonsinusoidal voltage and currents are present. *Displacement power factor* is the ratio of the fundamental watts to the fundamental volt-amperes, i.e.,

$$\text{PF}_D \triangleq \cos \theta_1 \qquad (7\text{-}38)$$

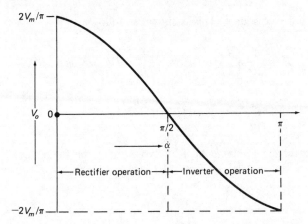

Fig. 7-20. V_o vs. α for Fig. 7-17, except just assuming continuous dc current.

where θ_1 is the phase angle between the fundamental components of voltage and current. The term *power factor* is the total power in watts divided by the product of the total RMS volts and the total RMS amperes, i.e.,

$$PF \triangleq \frac{\text{total power in watts}}{V_e I_e} \tag{7-39}$$

For the circuit in Fig. 7-17, these two power factors are quite easy to calculate, assuming negligible ripple in i_d. Using the results of Problem 15 in Chapter 1, the Fourier series expression for the square wave, with the $\omega t = 0$ axis at the positive-going zero crossing, is the following:

$$i_{ac}(\omega t) = \sum_{n=1}^{\infty} \frac{4I_d}{n\pi} \sin n\omega t \tag{7-40}$$

The v_1 waveform leads the square waveform of current by an angle α, as shown in Fig. 7-18 or Fig. 7-19. In accordance with Eq. (7-40), the positive-going zero crossing of the fundamental component of i_d occurs at the corresponding zero crossing of the square wave. This means that the displacement power factor for the circuit of Fig. 7-17 is cos α. Another way of showing this is to calculate the ratio of the fundamental watts to the fundamental volt-amperes. For the circuit of Fig. 7-17, where the voltage is a sine wave and the current is assumed to be a square wave, the fundamental watts are equal to the total watts delivered by the ac source. This is shown as follows

$$P = \frac{1}{2\pi} \int_0^{2\pi} vi\, d(\omega t) = \frac{1}{\pi} \int_\alpha^{\pi+\alpha} (V_m \sin \omega t) I_d\, d(\omega t)$$

$$= \frac{V_m I_d}{\pi} [-\cos \omega t]_\alpha^{\pi+\alpha} = \frac{2V_m I_d}{\pi} \cos \alpha \qquad (7\text{-}41)$$

The product of the fundamental RMS volts times the fundamental RMS amperes, where the voltage is sinusoidal and the current is a square wave, is as follows:

$$V_{1,e} I_{ac1,e} = \frac{V_m}{\sqrt{2}} I_{ac1,e} \qquad (7\text{-}42)$$

From Eq. (7-40), the amplitude of the fundamental component of i_{ac} is

$$I_{ac1,m} = \frac{4I_d}{\pi} \qquad (7\text{-}43)$$

The fundamental power is then given by

$$P_1 = \left(\frac{V_m}{\sqrt{2}}\right)\left(\frac{4I_d}{\sqrt{2}\,\pi}\right) \cos \alpha = \frac{2V_m I_d}{\pi} \cos \alpha \qquad (7\text{-}44)$$

This is the same as the total power given by Eq. (7-41). From Eqs. (7-41), (7-42), and (7-43), the displacement power factor is then

$$PF_D = \frac{(2V_m I_d/\pi) \cos \alpha}{(V_m/\sqrt{2})(4I_d/\sqrt{2}\,\pi)} = \cos \alpha \qquad (7\text{-}45)$$

The power factor in this case is

$$PF = \frac{(2V_m I_d/\pi) \cos \alpha}{(V_m/\sqrt{2})(I_d)} = \frac{2\sqrt{2}}{\pi} \cos \alpha \qquad (7\text{-}46)$$

The total power may also be determined by calculating the power delivered to the battery E, since all power taken from ac source v_1 in Fig. 7-17 must be delivered to the load battery. Thus,

$$P = (E)I_{d,\text{AVE}} = EI_d = \left(\frac{2V_m}{\pi} \cos \alpha\right) I_d = \frac{2V_m I_d}{\pi} \cos \alpha \qquad (7\text{-}47)$$

It is true in general that when a nonsinusoidal current is taken from a sinusoidal voltage source, the total power delivered is equal to the fundamental power delivered. This is shown as follows:

$$P = \frac{1}{2\pi} \int_0^{2\pi} vi\, d(\omega t)$$

$$= \frac{1}{2\pi} \int_0^{2\pi} [V_m \sin(\omega t + \alpha)]\left[\sum_{n=1}^{\infty} (A_n \cos n\omega t + B_m \sin n\omega t)\right] \qquad (7\text{-}48)$$

An arbitrary waveform of current is assumed to be flowing from a sinusoidal voltage source where the positive-going zero crossing of the periodic current waveform lags the sinusoidal voltage by an angle of α-degrees. The $t = 0$ axis is chosen as the positive-going zero crossing of the current, so the sinusoidal voltage leads this current by an angle of α-degrees.

$$P = \frac{V_m}{2\pi} \int_0^{2\pi} (\sin \omega t \cos \alpha + \cos \omega t \sin \alpha)$$

$$\times \left[\sum_{n=1}^{\infty} A_n \cos n\omega t + B_n \sin n\omega t\right] d(\omega t) \qquad (7\text{-}49)$$

From Eqs. (1-32) and (1-34) in Chapter 1, this reduces to

$$P = \frac{V_m}{2\pi} \int_0^{2\pi} (B_1 \cos \alpha \sin^2\omega t + A_1 \sin \alpha \cos^2\omega t)\, d(\omega t) \qquad (7\text{-}50)$$

Using Eq. (1-33), this yields

$$P = \frac{V_m}{2}[B_1 \cos \alpha + A_1 \sin \alpha] = \frac{V_m \sqrt{A_1^2 + B_1^2}}{2} \cos(\alpha - \beta) \qquad (7\text{-}51)$$

where $\beta = \tan^{-1}(A_1/B_1)$. The angle β is the angle by which the total fundamental component of the current leads the positive-going zero crossing of the current waveform. Thus, the fundamental power factor angle is $\alpha - \beta$.

7.2.6. Four Quadrant dc Motor Drives

Another very interesting and useful application of inverter operation of phase-controlled rectifiers is for regenerative braking in dc motor drives. Consider the circuit of Fig. 7-10. For simplicity, again it will be assumed that the motor has a separate fixed field and that there is negligible variation in the motor speed during a half-period of the power source frequency. Suppose that the motor is running at a constant speed with the conditions shown in Fig. 7-12. It is desired to stop the machine or to reverse rapidly the direction of rotation. Assume that a field-reversing controller is operated to reverse the direction of the field flux. This may take an appreciable amount of time, particularly for a large machine where the field time constant may be several seconds. However, for a relatively small machine—say, under 100 hp—the field flux can be reversed quite rapidly, so that it could be assumed that no significant change in speed took place during the field reversal. With this assumption, the motor counter-EMF should be essentially the same magnitude, but opposite in polarity, to that in Fig. 7-12 immediately after the field reversal. At the same time as the field is reversed, the gating angle must also be adjusted to call for the new desired steady-state speed. Suppose that the speed required is the same as for the previous operation, but with the opposite direction of rotation. For this case, the gating angle could be left at 60°, as in Fig. 7-12. However, generally it will need to be retarded to limit the motor current.

Figure 7-21 shows the waveforms immediately after the field flux has been reversed, assuming α is still 60° and negligible change in the motor speed during the field reversal. It is also assumed that the motor speed does not change

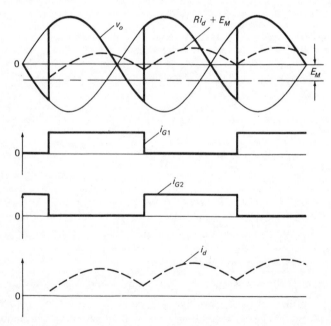

Fig. 7-21. Waveforms for Fig. 7-10 after field is reversed from that for Fig. 7-12.

significantly in the first few cycles after the field flux reversal. On subsequent cycles the motor speed would fall to zero and then build up to the final steady-state speed in the opposite direction. Since there is considerably greater motor current, if one compares Fig. 7-12 with Fig. 7-21, there is appreciable motor torque to cause a rapid speed reversal. The motor actually operates as a generator for the conditions in Fig. 7-21. This generator operation will continue until the CEMF has reversed polarity. It should be noted that the circuit is not operating as an inverter in Fig. 7-21. Rather, this situation is "plugging" where energy is taken both from the ac supply and the motor acting as a generator. Energy is not returned to the ac source, but it is dissipated in the armature circuit resistance.

The very high motor current can be avoided and energy can also be returned to the ac supply by operating the circuit of Fig. 7-10 as an inverter during the rapid speed reversal. Figure 7-22 shows the waveforms immediately after the field flux reversal, assuming negligible change in the motor speed during, and for the first few cycles shown after, the reversal. It is also assumed that α is reset equal to 120° to provide inverter operation. For this situation, it is apparent that the current will decrease until it becomes discontinuous. The average value of the voltage $v_o(t)$ must be equal to the average dc circuit voltage, $E_M + RI_d$, to maintain the dc current. In a practical situation, it would be desirable to advance the gating angle just enough on successive conduction intervals to maintain the average dc circuit current constant at the maximum safe value. This would rapidly bring the motor up to the desired speed in the reverse direction, at which point the waveforms would again be as in Fig. 7-12. All conditions would be the same, except the motor would be running in the reverse direction; but since the field also was reversed, the CEMF would be positive, as in Fig. 7-12.

Figure 7-22 also indicates a problem that may arise if the gating angles are retarded too far. The reapplication of forward voltage must not occur prior to the time required for the thyristor to recover its blocking capability. In the v_{SCR1} waveform of Fig. 7-22, it is noted that there is a significant interval of reverse voltage on SCR1 just after it has been conducting. If α were retarded too far, this reverse voltage interval for SCR1 would not be sufficient to guarantee turn-off. Thus, SCR1 would turn on again when v_1 became positive. This is a commutation failure. The load current is transferred only momentarily from SCR1 to SCR2. When v_1 becomes positive, causing SCR1 to come on again, SCR2 will go off. This may result in an excessively high current flow during this second consecutive conduction interval for SCR1.

One of the difficulties caused by pulse gating can be discerned as a result of further consideration of the waveforms in Figs. 7-11 through 7-22. If the motor CEMF is greater than the instantaneous ac supply voltage at the instant of gating, the next SCR to conduct will not come on. Then it will remain off for

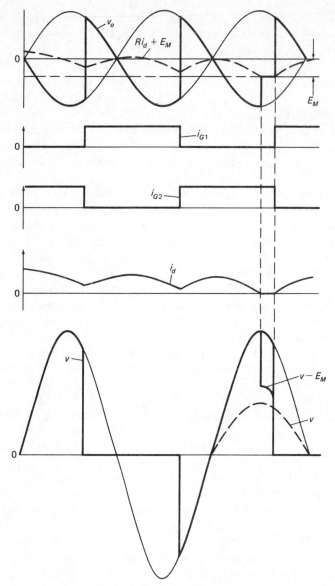

Fig. 7-22. Waveforms as in Fig. 7-21, except with $\alpha = 120°$.

the entire half-cycle of the power frequency, assuming that a gating pulse only in the order of 10 μs long is being used.

A four-quadrant dc motor drive can be produced using two circuits similar to those in Fig. 7-10. Figures 7-23 and 7-24 show two possible schemes. The term

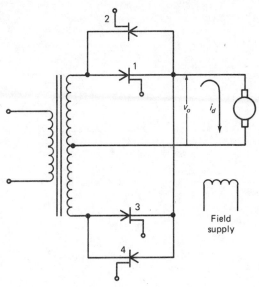

Fig. 7-23. Four-quadrant dc drive without circulating current limiting inductance.

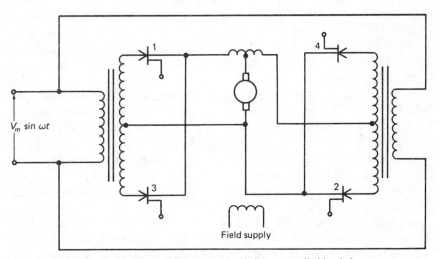

Fig. 7-24. Four-quadrant dc drive with circulating current limiting inductance.

four quadrant means that the phase-controlled rectifier/inverter can drive the motor in either direction, with rectifying or inverting operation in each direction. This is most simply illustrated by Fig. 7-25 for the circuit of Fig. 7-23. It should be noted that in this case, inverting operation occurs by reversing the

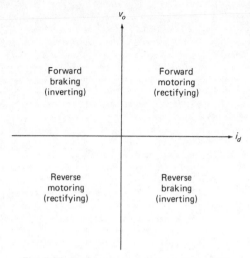

Fig. 7-25. Four-quadrant operation of Fig. 7-23.

current flow through the motor armature rather than by reversing the motor CEMF, which requires a field reversal. Thus, the field reversing is not required and much more rapid motor reversal is possible. In addition, these dc drives provide the fastest dynamic response to changes in torque or speed commands.

When the motor is being driven in one direction, SCR1 and SCR3 in Fig. 7-23 alternately conduct. A rapid reversal is obtained by inhibiting the gating of SCR1 and SCR3, and gating SCR2 and SCR4 on alternate half-cycles with a gating angle of greater than 90° so that the circuit including SCR2 and SCR4 operates in the inverting mode. A similar action occurs with the opposite pairs of SCRs for motoring in the other direction. For the circuit of Fig. 7-23, it is essential to sense the current and inhibit the firing of the even-numbered thyristors when positive motor current is flowing and the odd-numbered devices when negative motor current is flowing.

Figure 7-24 is a "circulating-current" type of circuit. The center-tapped inductor limits the current circulating between the two circuits, for example, when both SCR1 and SCR4 are conducting with the upper terminal of each transformer secondary positive. In both systems, the gating control should be arranged so that the circuit which is not conducting would produce the same average motor voltage if it were allowed to conduct as that being produced by the circuit which is supplying the motor voltage. This permits the most rapid transfer, with minimum load voltage transients, from rectifier to inverter operation or vice versa.

7.2.7. Other Single Phase Circuits

The second very basic single-phase-controlled rectifier circuit is the double-way or bridge circuit. Several different arrangements are shown in Fig. 7-26. The waveforms for Fig. 7-26(a) are essentially the same as for Fig. 7-10, except, of course, the peak reverse voltage across the thyristor is one-half the peak value for the thyristors in Fig. 7-10, assuming the same maximum dc output voltage. It is less expensive to use only two thyristors, as in Fig. 7-26(b). However, this modifies the control somewhat. The coasting diode in Fig. 7-26(c) prevents inverter operation of this circuit.

7.3. THREE-PHASE BRIDGE-CONTROLLED RECTIFIER

7.3.1. R and L-R Load

Figure 7-27 shows the very important three-phase bridge-controlled rectifier. Figure 7-28 illustrates the dc output voltage waveforms and required gating signals for a firing angle $\alpha = 15°$. Figure 7-29 shows the v_o and v_{SCR1} waveforms for $\alpha = 60°$. Figure 7-30 illustrates the v_o and $i_d R$ waveforms for $\alpha = 60°$, but assuming an L-R load where there is appreciable ripple in the dc current. The shaded areas again must be equal in Fig. 7-30 to satisfy the steady-state principle for the inductor. For both of these gating angles, the output voltage waveform is the same for either a purely resistive load, a purely inductive load, or any combination of L-R loads. It is important to note that α is again measured from the full-on point as a rectifier (that is, the point where conduction would begin if diodes were used instead of SCRs). In this case, the $\alpha = 0°$ point is not the zero crossing of the ac voltage waves but rather the intersection of a positive and a negative line-to-line voltage waveform.

Although it may seem a rather strange way to number the thyristors in Fig. 7-27, this system of numbering results in the sequence where the next higher number SCR is the next device to be gated in all cases. Of course, the sequence must be repeated so that SCR1 is the device to be gated following the gating of SCR6. Again the waveform of voltage across an SCR is of considerable importance. It must be positive at the instant of gating in order for the device being gated to start conducting. On the other hand, it must be negative at the moment it is commutated off by gating the next device.

An SCR voltage waveform is easily determined by the following procedure. Suppose the waveform desired is v_{SCR1}. First, note the SCRs which are conducting during each time interval. Then, find the particular SCR that is conducting and is connected to the same dc output line. For the first two 60° intervals in Fig. 7-29, SCR5 is the device conducting that is connected to the same dc

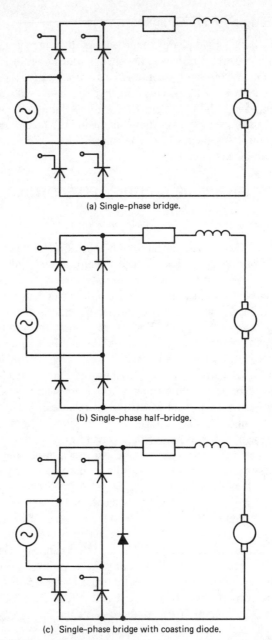

(a) Single-phase bridge.

(b) Single-phase half-bridge.

(c) Single-phase bridge with coasting diode.

Fig. 7-26. Single-phase bridge-controlled rectifier circuits. (a) Single-phase bridge; (b) single-phase half-bridge; (c) single-phase bridge with coasting diode.

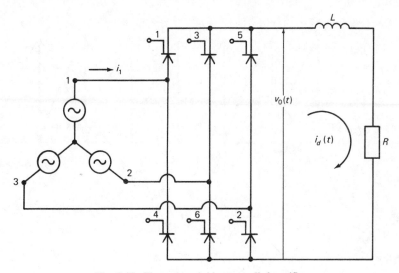

Fig. 7-27. Three-phase bridge-controlled rectifier.

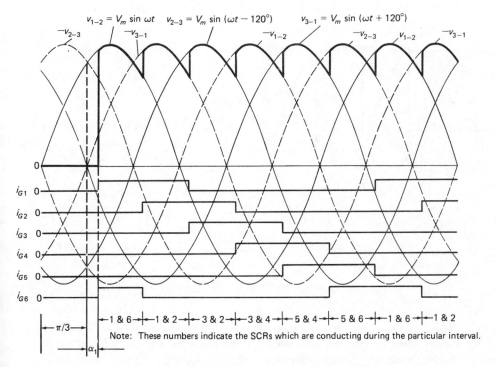

Note: These numbers indicate the SCRs which are conducting during the particular interval.

Fig. 7-28. $v_o(t)$ for circuit in Fig. 7-27 with $\alpha = 15°$.

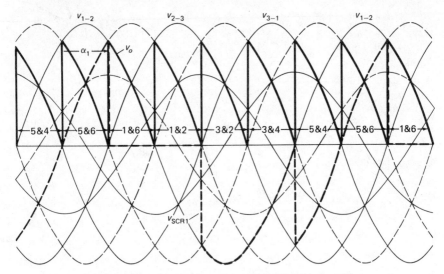

Fig. 7-29. v_o and v_{SCR1} for circuit in Fig. 7-27 with $\alpha = 60°$.

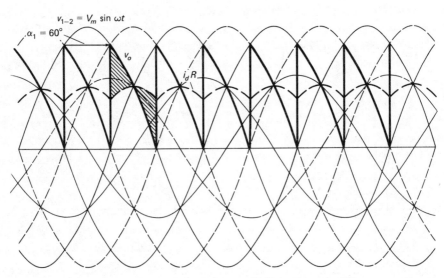

Fig. 7-30. v_o and $i_d R$ for Fig. 7-27, with ripple in i_d ($\alpha = 60°$).

output line as SCR1. Finally, note that the voltage across the SCR is the line-to-line voltage connected between the SCR conducting and the SCR for which the anode-cathode voltage waveform is desired. During the first two 60° intervals in Fig. 7-29, v_{1-3} is the line-to-line voltage connected between the anodes of SCR1 and SCR5. Therefore, v_{SCR1} is equal to v_{1-3} during these two 60° intervals. This same procedure is used for all subsequent time intervals, except, of course, the SCR voltage is zero (neglecting forward drop) during its conduction intervals.

When α is greater than 60°, discontinuous conduction may occur. Thus, the v_o waveform may not be the same for both a purely resistive load and an L-R load. Figure 7-31 illustrates the v_o waveform for $\alpha = 90°$ with a resistive load. Figure 7-32 shows the v_o and $i_d R$ waveforms for $\alpha = 90°$ with an inductive load, where there is appreciable ripple in i_d. In both of these cases, the dc current is discontinuous.

7.3.2. CEMF Load and Inverter Operation

When the dc current is continuous, the expression for the average dc output voltage for the circuit in Fig. 7-27 is derived as follows:

$$V_o = \frac{1}{2\pi} \int_0^{2\pi} v_o(\omega t)\, d(\omega t) = \frac{1}{\pi/3} \int_{(\pi/3)+\alpha}^{(2\pi/3)+\alpha} V_m \sin \omega t\, d(\omega t)$$

$$= \frac{3V_m}{\pi}[-\cos \omega t]\Big|_{(\pi/3)+\alpha}^{(2\pi/3)+\alpha} = \frac{3V_m}{\pi} \cos \alpha \qquad (7\text{-}52)$$

As shown in Fig. 7-30, the period of the v_o waveform is one-sixth of the power-frequency period. The particular period used to obtain Eq. (7-52) is the one

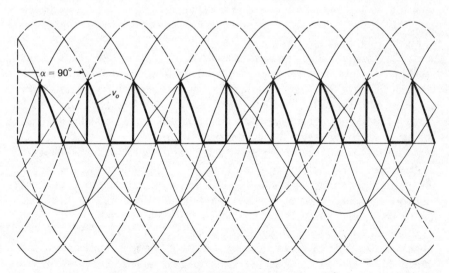

Fig. 7-31. v_o for Fig. 7-27 with R-load and $\alpha = 90°$.

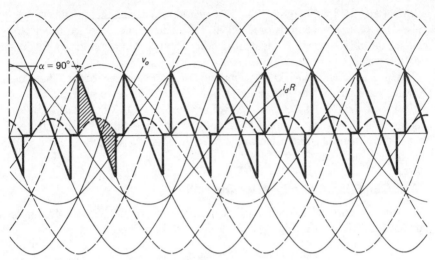

Fig. 7-32. v_o and $i_d R$ for Fig. 7-27, with ripple in i_d ($\alpha = 90°$).

starting when SCR1 is first gated. During this first conduction interval for SCR1, the dc voltage is following v_{1-2}. This first conduction interval for SCR1 is from $\pi/3$ to $(2\pi/3) + \alpha$.

Equation (7-52) is the same cosine curve as shown in Fig. 7-20, but the amplitude is now $3V_m/\pi$ instead of $2V_m/\pi$. Again, this characteristic assumes continuous dc circuit current. Also, steady-state inverter operation can occur only if there is a source of power in the dc circuit, such as a negative battery or CEMF.

Figure 7-33 illustrates the waveforms for Fig. 7-27, except with a dc motor load. The complete load circuit is assumed to be a series R-L-CEMF, which would be the case for a separately excited dc motor. The CEMF is essentially constant in Fig. 7-33, which implies that there is negligible variation in the motor speed during 60° of the power-frequency cycle. Again, it is possible to brake the motor regeneratively by reversing the field excitation and then adjusting the phase-control angle for inverter operation.

Figure 7-34 is a full reversing circuit referred to as a four-quadrant dc drive. In this case, it is possible to drive the motor in either direction and to feed energy from the motor back into the ac power supply to achieve regenerative fast reversal and/or braking. The separate field excitation remains fixed, and the reversing and/or regenerative action is achieved by inhibiting the firing of the rectifying bridge and gating the other bridge to operate in the inverting mode. It should be noted that very reliable gating logic and control is required. It is possible to short the ac lines through thyristors in the two different

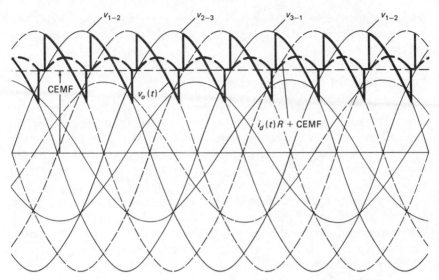

Fig. 7-33. Waveforms for Fig. 7-27, except with dc motor load (R-L-CEMF).

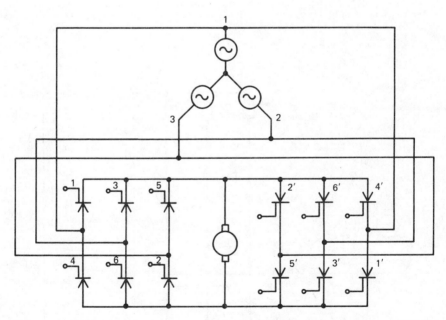

Note: Assume separately excited field.

Fig. 7-34. Four-quadrant dc drive.

bridges during reversing or regenerative braking operations. There are also both "circulating-current" and "non-circulating-current" polyphase four-quadrant converter circuits.

7.3.3. Effects of ac Line Reactance

Figure 7-35 includes three identical inductors in series with the ac lines to represent the combined effects of power source, line, and input transformer reactance. The effects on the operation of the circuit are indicated in Fig. 7-36. For simplicity, it is assumed that there is negligible ripple in the dc current. In addition, the gating angle is changed from one conducting interval to the next to show how the commutating angle μ varies as a function of the gating angle α.

An expression for the commutating angle can be derived quite easily with the assumption of negligible variation in the dc current during commutation. Consider the particular commutation in Fig. 7-36 when SCR1 is gated to transfer current from SCR5 to SCR1. During this overlap interval, both SCR1 and SCR5 are conducting. Thus,

$$v_{1-3} = v_{LC1} + v_{LC3} \tag{7-53}$$

Also,

$$i_{SCR1} + i_{SCR5} = I_{d\mu} \tag{7-54}$$

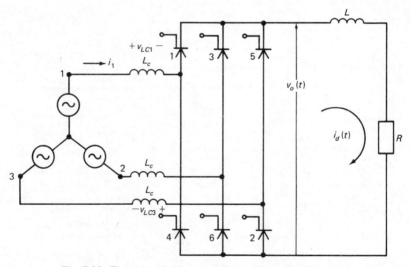

Fig. 7-35. Three-phase bridge-controlled rectifier with ac reactance.

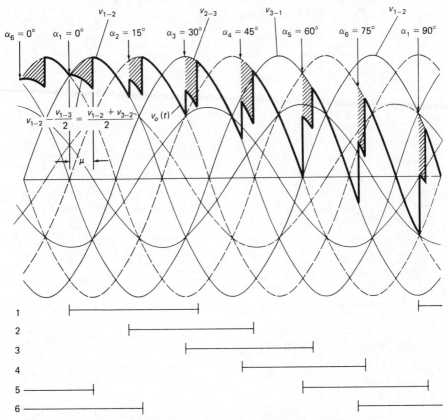

Note: Shaded areas are equal assuming negligible ripple in $i_d(t)$
and thus essentially constant dc

Fig. 7-36. $v_o(t)$ and SCR conduction periods for circuit in Fig. 7-35 with varying α.

where $I_{d\mu}$ is essentially constant. Thus,

$$\frac{di_{SCR1}}{dt} + \frac{di_{SCR2}}{dt} = 0 \qquad (7\text{-}55)$$

Since the commutating inductances are assumed equal, Eq. (7-55) can be used in Eq. (7-53) to give

$$v_{LC} = \frac{v_{1-3}}{2} \qquad (7\text{-}56)$$

Also, during the commutating angle μ,

$$v_o = v_{1-2} - v_{LC} = v_{1-2} - \frac{v_{1-3}}{2} = v_{1-2} - \frac{(v_{1-2} - v_{3-2})}{2} = \frac{v_{1-2} + v_{3-2}}{2}$$

$$(7\text{-}57)$$

(Note: There is negligible change in the current through inductance L_c in series with line 2, because the essentially constant current I_d is flowing through SCR6 back into this ac line.) In this case, the output voltage v_o is not zero during the commutation interval as it was in the single-phase case, assuming negligible change in the dc output current during commutation. However, v_o is the instantaneous average of v_{1-2} and v_{3-2}, or it is halfway between the line-to-line voltage which determines v_o immediately before commutation and the line-to-line voltage which establishes v_o immediately after commutation. This general rule can be used for all commutations, assuming negligible change in i_d during commutation, unless the ac circuit inductance is so large that a particular commutation is not over before the next device to conduct is gated.

For the specific commutation interval where SCR1 is beginning to conduct and SCR5 is being commutated off,

$$i_{\text{SCR1}}(\theta) = \frac{1}{\omega L_c} \int_{(\pi/3)+\alpha}^{(\pi/3)+\alpha+\mu} \frac{v_{1-3}}{2} \, d(\omega t) \qquad (7\text{-}58)$$

or

$$i_{\text{SCR1}(\mu)} = I_{d\mu} = \frac{1}{2\omega L_c} \int_{(\pi/3)+\alpha}^{(\pi/3)+\alpha+\mu} V_m \sin(\omega t - \pi/3) \, d(\omega t)$$

$$= \frac{V_m}{2\omega L_c} [-\cos(\omega t - \pi/3)] \Big|_{(\pi/3)+\alpha}^{(\pi/3)+\alpha+\mu} = \frac{V_m}{2\omega L_c} [\cos \alpha - \cos(\alpha + \mu)]$$

$$(7\text{-}59)$$

and

$$\cos(\alpha + \mu) = \cos \alpha - \frac{2\omega L_c I_{d\mu}}{V_m} \qquad (7\text{-}60)$$

7.4. ADDITIONAL REMARKS

There are a tremendous variety of phase-controlled rectifiers arrangements included in the books listed in the references at the end of this chapter. Two basic circuits have been discussed here — the single-phase, center-tap-controlled rectifier and the three-phase bridge-controlled rectifier. It is hoped that this chapter has provided the foundation preparing the student to cope with a wide range of single- and polyphase-controlled rectifiers.

REFERENCES

1. Marti, O. K., and Winograd, H. *Mercury Arc Power Rectifiers*. New York: McGraw–Hill, 1930.
2. Rissik, H. *The Fundamental Theory of Arc Converters*. New York: Chapman and Hall, 1939.
3. Prince, D. C., and Vodges, F. B. *Principles of Mercury Arc Rectifiers and Their Circuits*. New York: McGraw–Hill, 1947.
4. Bedford, B. D., and Hoft, R. G. *Principles of Inverter Circuits*. New York: John Wiley & Sons, 1964.
5. Schafer, J. *Rectifier Circuits, Theory and Design*. New York: John Wiley & Sons, 1965.
6. Pelly, B. R. *Thyristor Phase-Controlled Converters and Cycloconverters*. New York: Wiley, 1971.
7. Scoles, Graham J. *Handbook of Rectifier Circuits*. New York: Halsted Press-Division of John Wiley & Sons, 1980.

PROBLEMS

1. Carefully sketch possible waveforms for v_o and i_d in Fig. 7-1, with $v_1 = 200 \sin \omega t$, a 100-V battery added in series with the L_d-R load such that it is being charged, and assuming that L_d and R are such that the dc current is discontinuous.

2. Consider the phase-controlled rectifier of Fig. 7-5.
   a) Derive the general expression for V_o as a function of α and V_m.
   b) Carefully sketch the waveforms for v_o and i_{ac} with $\alpha = 60°$.
   c) With $v_1 = \sqrt{2}\, 220 \sin \omega t$, $R = 20\ \Omega$, and $\alpha = 60°$, calculate the power delivered to the load resistor and the power factor at the ac source.

3. Consider the single-phase-controlled rectifier of Fig. 7-7, assuming negligible ripple in the dc current i_d, $v_1 = \sqrt{2}\, 220 \sin \omega t$, $R = 20\ \Omega$, and $\alpha = 45°$.
   a) Carefully sketch the steady-state waveforms for v_o, i_{ac}, v_{SCR1}, and i_{SCR1}.
   b) Determine the average value of the dc current I_d.
   c) Determine the power factor at the ac source.

4. Consider the circuits in Fig. 7-26. Carefully sketch possible steady-state waveforms for v_o, i_{ac}, and v_{SCR1} (upper left SCR), assuming the following:
   a) Negligible ripple in i_d, zero motor CEMF, and $\alpha = 45°$.
   b) A motor CEMF equal to one-half the ac source voltage amplitude, $\alpha = 90°$, and an L_d-R combination such that the current is discontinuous.

5. Consider the single-phase bridge-controlled rectifier of Fig. 7-26(a), except with zero CEMF and an inductance L_c directly in series with the ac source.

Assuming sufficient L_d in the dc circuit such that there is negligible ripple in i_d, $\alpha = 45°$, and a commutation angle of about $15°$, carefully sketch v_o, i_{ac}, v_{SCR1} (upper left SCR), and v_{Lc}.

6. Consider the circuit of Fig. 7-17, with the addition of a resistance $R = 2\ \Omega$ directly in series with the load battery equal negative 125 V (E is positive at the transformer center tap). Assuming $I_d = 20$ A with negligible ripple and $v_1 = 220\sqrt{2} \sin \omega t$,
   a) Calculate the gating angle α for steady-state operation.
   b) Carefully sketch steady-state waveforms for v_o, i_{ac}, and v_{SCR1}.
   c) Determine the power factor at the ac source.

7. Consider the circuit of Fig. 7-17 again, except with the inductor L_d replaced by a 2-Ω resistor, and the battery voltage $E = -100$ V, $v_1 = 300 \sin \omega t$, and $\alpha = 120°$.
   a) Carefully sketch waveforms for v_o, i_d, and i_{ac}.
   b) Calculate the power delivered by the dc source.

8. Consider the three-phase bridge-controlled rectifier of Fig. 7-27, assuming negligible ripple in I_d and $\alpha = 75°$.
   a) Carefully sketch the steady-state waveforms for v_o, i_1, and v_{SCR3}.
   b) Calculate the power factor at the ac source with $v_{1-2} = 460\sqrt{2} \sin \omega t$ and $R = 2\ \Omega$.

9. Consider the circuit of Fig. 7-27 except with SCR4, SCR6, and SCR2 replaced with diodes. Assume negligible ripple in the dc current and $\alpha = 75°$. Carefully sketch the steady-state waveforms for v_o, v_{SCR1}, and i_1.

10. Consider the three-pulse (three-phase, single-way) circuit similar to Fig. 7-27. The three SCRs connected to the negative dc output line are removed, and this negative line is connected to the neutral of the ac source. Assume negligible ripple in the dc current and $\alpha = 60°$. Carefully sketch the steady-state waveforms for v_o, v_{SCR1}, and i_1.

11. Consider Fig. 7-27 but include a 300-V battery connected in series with the resistance $R = 0.2\ \Omega$ and with the positive battery terminal connected to the anodes of SCR2, SCR4, and SCR6. Assume negligible ripple in the dc current, $I_d = 100$ A, and $v_{1-2} = 440\sqrt{2} \sin \omega t$.
   a) Determine the gating angle α for steady-state operation.
   b) Carefully sketch the steady-state waveforms for v_{SCR1}, v_o, and i_1.
   c) Calculate the power factor at the ac source.
   d) Calculate the displacement power factor at the ac source.

12. Consider the circuit of Fig. 7-35 but replace the resistor R by a battery with its positive terminal connected to the lower dc line. Assume negligible ripple in i_d, steady-state operation with $\alpha = 160°$, and a commutation angle of $10°$. Starting with the instant when α_1 is changed to $175°$, carefully sketch v_o, v_{Lc}, v_{SCR1}, and i_1 for two complete cycles of the supply frequency to show the effect of the commutation failure.

8
CYCLOCONVERTERS

8.1. INTRODUCTION

One of the most complex power electronic circuits is the cycloconverter. The three-phase arrangement that is most widely used involves 36 reverse-blocking triode (SCR) thyristors. Thus, the number of power and control circuit components is large, and analysis of the operating waveforms is quite involved. A serious study of references [1] through [4] is recommended for one interested in cycloconverter analysis and design. The object of this chapter is to present the basic principles of the cycloconverter and then to illustrate the harmonic analysis method described by Pelly [1].

8.2. BASIC PRINCIPLES

The cycloconverter is essentially a pair of phase-controlled rectifiers connected to a common load. One rectifier produces a positive voltage across the load, and the second supplies negative load voltage. If the two phase-controlled rectifiers are alternately turned on and off, an alternating voltage is produced at the load. Thus, normally the cycloconverter supplies an ac output at a lower frequency than the power source.

Figure 8-1 illustrates two versions of one of the simplest cycloconverters. It should be noted that the two circuits in Fig. 8-1 are the same as the four-quadrant converters of Chapter 7 (Figs. 7-23 and 7-24). However, the four-quadrant converter is controlled in an aperiodic fashion, whereas the cycloconverter is controlled to supply periodic ac output to the load.

Figure 8-2 shows several possible output-voltage waveforms with resistive load. These are not of practical significance, but they do illustrate the most elementary operating modes. They are not of practical use because the load is almost never purely resistive and the harmonic content in the waveforms of Fig. 8-2 is large. As an illustration of this latter point, the harmonic content of Fig. 8-2(b) is determined. Since this waveform is an odd function, with $\omega t = 0$ chosen as the starting point of Fig. 8-2(b), no cosine terms appear in the Fourier series. Also with this choice for the $\omega t = 0$ axis, the waveform of Fig. 8-2(b) has half-wave symmetry (Chapter 1, Section 1.6) so that no even harmonics are present.

The amplitudes of the sinusoidal terms in the Fourier series for Fig. 8-2(b) are calculated from Eq. (1-30) as follows:

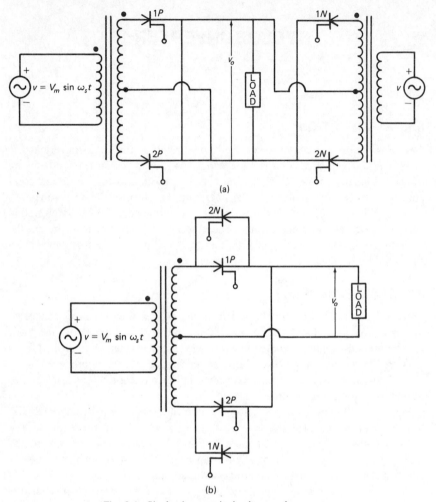

Fig. 8-1. Single-phase-to-single-phase cycloconverter.

$$B_n = \frac{1}{\pi} \int_0^{2\pi} f(\omega_o t) \sin n\omega_o t \, d(\omega_o t)$$

$$= \frac{1}{\pi} \left[\int_0^{\pi/3} V_m \sin 3\omega_o t \, \sin n\omega_o t \, d(\omega_o t) - \int_{\pi/3}^{2\pi/3} V_m \sin 3\omega_o t \, \sin n\omega_o t \, d(\omega_o t) \right.$$

$$\left. + \int_{2\pi/3}^{4\pi/3} V_m \sin 3\omega_o t \, \sin n\omega_o t \, d(\omega_o t) \right.$$

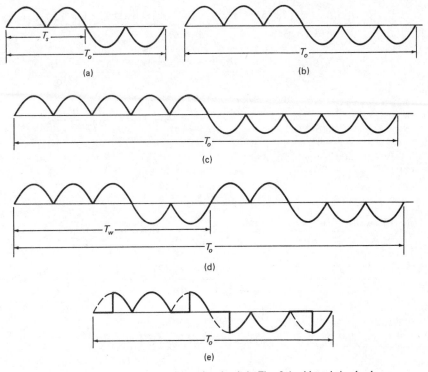

Fig. 8-2. Elementary waveforms for circuit in Fig. 8-1 with resistive load.

$$-\int_{4\pi/3}^{5\pi/3} V_m \sin 3\omega_o t \, \sin n\omega_o t \, d(\omega_o t)$$

$$+\int_{5\pi/3}^{2\pi} V_m \sin 3\omega_o t \, \sin n\omega_o t \, d(\omega_o t) \Bigg]$$ (8-1)

If n is not equal to 3, then

$$B_n = \frac{V_m}{2\pi} \left[\frac{\sin(3-n)\omega_o t}{(3-n)} - \frac{\sin(3+n)\omega_o t}{(3+n)} \right] \Bigg|_0^{\pi/3} (-) \Bigg|_{\pi/3}^{2\pi/3} (+) \Bigg|_{2\pi/3}^{4\pi/3} (-) \Bigg|_{4\pi/3}^{5\pi/3} (+) \Bigg|_{5\pi/3}^{2\pi}$$

$$= \frac{V_m}{2\pi} \left[\frac{\sin(3-n)(\pi/3)}{(3-n)} - \frac{\sin(3+n)(\pi/3)}{(3+n)} - \frac{\sin(3-n)(2\pi/3)}{(3-n)} \right.$$

$$\left. + \frac{\sin(3+n)(2\pi/3)}{(3+n)} + \frac{\sin(3-n)(\pi/3)}{(3-n)} - \frac{\sin(3+n)(\pi/3)}{(3+n)} \right.$$

$$+ \frac{\sin(3 - n)(4\pi/3)}{(3 - n)} - \frac{\sin(3 + n)(4\pi/3)}{(3 + n)} - \frac{\sin(3 - n)(2\pi/3)}{(3 - n)}$$

$$+ \frac{\sin(3 + n)(2\pi/3)}{(3 + n)} - \frac{\sin(3 - n)(5\pi/3)}{(3 - n)} + \frac{\sin(3 + n)(5\pi/3)}{(3 + n)}$$

$$+ \frac{\sin(3 - n)(4\pi/3)}{(3 - n)} - \frac{\sin(3 + n)(4\pi/3)}{(3 + n)} - \frac{\sin(3 - n)(5\pi/3)}{(3 - n)}$$

$$+ \left. \frac{\sin(3 + n)(5\pi/3)}{(3 + n)} \right]$$

$$= \frac{V_m}{\pi} \left[\frac{\sin(3 - n)(\pi/3)}{(3 - n)} - \frac{\sin(3 - n)(2\pi/3)}{(3 - n)} + \frac{\sin(3 - n)(4\pi/3)}{(3 - n)} \right.$$

$$- \frac{\sin(3 - n)(5\pi/3)}{(3 - n)} - \frac{\sin(3 + n)(\pi/3)}{(3 + n)} + \frac{\sin(3 + n)(2\pi/3)}{(3 + n)}$$

$$- \left. \frac{\sin(3 + n)(4\pi/3)}{(3 + n)} + \frac{\sin(3 + n)(5\pi/3)}{(3 + n)} \right]$$

$$= \frac{V_m}{\pi} \left\{ \left(\sin\left[\pi - \left(\frac{n\pi}{3} \right) \right] - \sin\left[2\pi - \left(\frac{2n\pi}{3} \right) \right] + \sin\left[4\pi - \left(\frac{4n\pi}{3} \right) \right] \right. \right.$$

$$- \left. \sin\left[5\pi - \left(\frac{5n\pi}{3} \right) \right] \right/ (3 - n) \Bigg)$$

$$+ \left(-\sin\left[\pi + \left(\frac{n\pi}{3} \right) \right] + \sin\left[2\pi + \left(\frac{2n\pi}{3} \right) \right] \right.$$

$$- \left. \sin\left[4\pi + \left(\frac{4n\pi}{3} \right) \right] + \sin\left[5\pi + \left(\frac{5n\pi}{3} \right) \right] \right/ (3 + n) \Bigg) \Bigg\}$$

$$= \frac{V_m}{\pi} \left[\frac{\sin(n\pi/3) + \sin(2n\pi/3) - \sin(4n\pi/3) - \sin(5n\pi/3)}{(3 - n)} \right.$$

$$+ \left. \frac{\sin(n\pi/3) + \sin(2n\pi/3) - \sin(4n\pi/3) - \sin(5n\pi/3)}{(3 + n)} \right]$$

$$= \frac{2V_m}{\pi} \left[\frac{\sin(n\pi/3) + \sin(2n\pi/3)}{(3 - n)} + \frac{\sin(n\pi/3) + \sin(2n\pi/3)}{(3 + n)} \right]$$

$$B_n = \frac{12V_m}{\pi} \left\{ \frac{\sin(n\pi/3) + \sin[n\pi - (n\pi/3)]}{9 - n^2} \right\} = \frac{24V_m}{(9 - n^2)\pi} \sin \frac{n\pi}{3} \qquad (8\text{-}2)$$

When $n = 3$, then

$$B_3 = \frac{V_m}{\pi} \left[\int_0^{\pi/3} \sin^2 3\omega_o t \, d(\omega_o t) - \int_{\pi/3}^{2\pi/3} \sin^2 3\omega_o t \, d(\omega_o t) + \ldots \right]$$

$$= \frac{V_m}{2\pi} \left[\omega_o t - \frac{\sin 6\omega_o t}{6} \right] \Big|_0^{\pi/3} (-) \Big|_{\pi/3}^{2\pi/3} (+) \Big|_{2\pi/3}^{4\pi/3} (-) \Big|_{4\pi/3}^{5\pi/3} (+) \Big|_{5\pi/3}^{2\pi}$$

$$= \frac{V_m}{2\pi} \left[\frac{\pi}{3} - \left(\frac{2\pi}{3} - \frac{\pi}{3} \right) + \left(\frac{4\pi}{3} - \frac{2\pi}{3} \right) - \left(\frac{5\pi}{3} - \frac{4\pi}{3} \right) + \left(2\pi - \frac{5\pi}{3} \right) \right.$$

$$- \frac{\sin 2\pi}{6} + \frac{\sin 4\pi}{6} - \frac{\sin 2\pi}{6} - \frac{\sin 8\pi}{6} + \frac{\sin 4\pi}{6} + \frac{\sin 10\pi}{6}$$

$$\left. - \frac{\sin 8\pi}{6} - \frac{\sin 12\pi}{6} + \frac{\sin 10\pi}{6} \right]$$

$$= \frac{V_m}{2\pi} \left[\frac{\pi}{3} - \frac{\pi}{3} + \frac{2\pi}{3} - \frac{\pi}{3} + \frac{\pi}{3} \right] = \frac{V_m}{3} \qquad (8\text{-}3)$$

Thus, from Eqs. (8-2) and (8-3) the Fourier series for the waveform of Fig. 8-2(b) is

$$v_o(\omega_o t) = \frac{3\sqrt{3} \, V_m}{2\pi} \sin \omega_o t + \frac{V_m}{3} \sin 3\omega_o t + \frac{3\sqrt{3} \, V_m}{4\pi} \sin 5\omega_o t$$

$$- \frac{3\sqrt{3} \, V_m}{10\pi} \sin 7\omega_o t + \frac{3\sqrt{3} \, V_m}{28\pi} \sin 11\omega_o t - \frac{3\sqrt{3} \, V_m}{40\pi} \sin 13\omega_o t$$

$$+ \frac{3\sqrt{3} \, V_m}{70\pi} \sin 17\omega_o t - \frac{3\sqrt{3} \, V_m}{88\pi} \sin 19\omega_o t + \ldots \qquad (8\text{-}4)$$

or

$$v_o(\omega_o t) = \frac{3\sqrt{3}\, V_m}{2\pi}\left[\sin \omega_o t + \frac{2\pi}{9\sqrt{3}} \sin 3\omega_o t + \frac{1}{2} \sin 5\omega_o t - \frac{1}{5} \sin 7\omega_o t \right.$$

$$+ \frac{1}{14} \sin 11\omega_o t - \frac{1}{20} \sin 13\omega_o t + \frac{1}{35} \sin 17\omega_o t$$

$$\left. - \frac{1}{44} \sin 19\omega_o t + \ldots \right] \qquad (8\text{-}5)$$

The harmonic content of the waveform of Fig. 8-2(b) expressed as a percentage of the fundamental is then the following:

n	$\left\|\dfrac{B_n}{B_1}\right\| \times 100\%$
1	100%
3	40.3%
5	50%
7	20%
9	0
11	7.1%
13	5%
15	0
17	2.9%
19	2.3%
21	0
23	1.5%
.	.

$$n \qquad \frac{16}{(9 - n^2)\sqrt{3}} \left(\sin \frac{n\pi}{3} \right) \times 100\%$$

Although the fundamental component in the output waveform is the largest, there is about 40 percent third harmonic, 50 percent fifth harmonic, and 20 percent seventh harmonic. The higher order harmonics reduce in amplitude, and there are no multiples of the third harmonic. However, in a practical situation the harmonic content of the waveform generally would not be acceptable.

Figure 8-2(d) illustrates another mode of operation. For waveforms of the type shown in Fig. 8-2(a) through (c), the output period is always an integer multiple of the input period.

$$T_o = nT_s \tag{8-6}$$

or

$$f_o = \frac{f_s}{n} \tag{8-7}$$

where n is any positive integer, T_o is the period of the output waveform, and T_s is the supply period. Of course, this limits the number of possible output frequencies. Waveforms of the type illustrated in Fig. 8-2(d) permit additional flexibility in the frequency ratio, but the harmonic content is still large, and subharmonics of the desired or wanted output are also present. The general relationship for waveforms like Fig. 8-2(d) is

$$T_w = n\frac{T_s}{2} \tag{8-8}$$

or

$$f_w = \frac{f_s}{n/2} \tag{8-9}$$

where again n is any positive integer and T_s is the supply period, but now T_w is the period of the "wanted" or desired output waveform, i.e., the period of the fundamental component of the desired output. The Fourier series representing

Fig. 8-2(d) is (see Problem 1)

$$v(\omega_o t) = \frac{20V_m}{\pi}[.0396 \sin \omega_o t + .1186 \sin 2\omega_o t - .0367 \sin 3\omega_o t$$

$$+ .0249 \sin 4\omega_o t + .0314 \sin 5\omega_o t + .0204 \sin 6\omega_o t$$

$$- .0245 \sin 7\omega_o t + .0638 \sin 8\omega_o t + .017 \sin 9\omega_o t$$

$$- .0099 \sin 11\omega_o t + \ldots] \tag{8-10}$$

Equation (8-10) is obtained using a similar procedure to that required for the derivation of Eq. (8-5). The wanted component is the second harmonic of ω_o in this case. The harmonic content of v_o expressed as a percentage of the wanted component is as follows:

n	f	% of f_w Component
1	$(1/2)f_w$	33.4
2	f_w	100.
3	$(3/2)f_w$	30.9
4	$2f_w$	21.
5	$(5/2)f_w$	8.4
6	$3f_w$	17.2
7	$(7/2)f_w$	46.2
8	$4f_w$	53.8
9	$(9/2)f_w$	14.3
10	$5f_w$	0
11	$(11/2)f_w$	8.3
$\cdot$	$\cdot$	$\cdot$
$\cdot$	$\cdot$	$\cdot$
$\cdot$	$\cdot$	$\cdot$

Thus, for the waveform of Fig. 8-2(d), the harmonic content is again quite large. Another very disadvantageous feature of this waveform is the presence of the large subharmonic component at one-half the wanted output frequency. Generally it is quite difficult to filter subharmonics. It is theoretically possible to control the gating instants to suppress subharmonics, but this only works well when the subharmonic is at least several to one below the wanted frequency.

Figure 8-2(e) illustrates a technique which can be used to reduce the unwanted components in the cycloconverter output voltage waveform. This is referred to here as "programmed" gating. The idea is to change the gating angle on subsequent half-cycles of the supply waveform so that the desired output frequency component is maximized. The harmonic components, up to the fifteenth, for the waveform in Fig. 8-2(e) are (see Problem 2) as follows:

n	A_n	B_n	$\sqrt{A_n^2 + B_n^2}$	$\dfrac{\sqrt{A_n^2 + B_n^2}}{\sqrt{A_1^2 + B_1^2}} \times 100\%$
1	$\dfrac{-V_m}{8\pi}$	$\dfrac{9\sqrt{3}\,V_m}{8\pi}$	$\dfrac{\sqrt{61}\,V_m}{4\pi}$	100%
3	$\dfrac{-2V_m}{3\pi}$	0	$\dfrac{2V_m}{3\pi}$	34%
5	$\dfrac{-7V_m}{16\pi}$	$\dfrac{9\sqrt{3}\,V_m}{16\pi}$	$\dfrac{\sqrt{73}\,V_m}{8\pi}$	54.7%
7	$\dfrac{17V_m}{40\pi}$	$\dfrac{-9\sqrt{3}\,V_m}{40\pi}$	$\dfrac{\sqrt{133}\,V_m}{20\pi}$	29.5%
9	$\dfrac{2V_m}{3\pi}$	0	$\dfrac{2V_m}{3\pi}$	34%
11	$\dfrac{25V_m}{112\pi}$	$\dfrac{9\sqrt{3}\,V_m}{112\pi}$	$\dfrac{\sqrt{217}\,V_m}{56\pi}$	13.5%
13	$\dfrac{-23V_m}{160\pi}$	$\dfrac{-9\sqrt{3}\,V_m}{160\pi}$	$\dfrac{\sqrt{193}\,V_m}{80\pi}$	8.9%
15	$\dfrac{-2V_m}{9\pi}$	0	$\dfrac{2V_m}{9\pi}$	11.4%

Unfortunately, the harmonic content is not significantly changed for the waveform in Fig. 8-2(e) compared to that for Fig. 8-2(b). In this case, programmed gating did not accomplish its goal. However, this gating method works much better when the ratio between the supply frequency and the output frequency is large. This can be explained in a qualitative fashion by a brief discussion of the cosine wave crossing (CWC) programmed gating method ([5], [6]).

The principle of the CWC gating method is that during each half-cycle of the supply, the appropriate thyristor is gated at the instant which will produce an average voltage over the half-cycle interval equal to the instantaneous voltage of the wanted output sinusoidal wave during the same interval. Of course, this is not strictly possible since the average over an interval — which is a constant for the interval — cannot equal the instantaneous value of a function over that interval because the instantaneous value changes during the interval. However, as the supply frequency becomes large relative to the wanted output frequency, in the limit the half-cycle period of the supply becomes vanishingly small compared to the output period. Thus, the wanted output waveform over the very short time interval is approximately constant so that it can be nearly equal to the average voltage from the phase-controlled rectifier for the particular supply half-period.

Figure 8-3 illustrates the CWC programmed gating method for the single-phase cycloconverter of Fig. 8-1. The $v_{o,P}$ and $v_{o,N}$ waveforms are drawn assuming continuous current where

$$V_o = \frac{2V_{s,m}}{\pi} \cos \alpha \qquad (8\text{-}11)$$

if α is constant. Actually, α varies from one power supply half-cycle to the next, as shown in Fig. 8-3. On the positive half-cycle of the reference or wanted output sinewave,

$$v_{o,w} = V_{o,m} \sin \omega_w t \qquad (8\text{-}12)$$

the output voltage is the $v_{o,P}$ waveform. Conversely, $v_{o,N}$ is the output waveform on the negative half-cycle of Eq. (8-12). In practice, the gating control is designed such that the gate signals for both the positive and negative phase-controlled rectifiers are always being generated. However, only one of the two rectifiers is gated during each half-period of the $v_{o,w}$ waveform. Functionally, this can be accomplished by inhibiting the gating of one phase-controlled rectifier on the appropriate half-cycle of the wanted output waveform. For example, when $v_{o,w}$ is positive, the gate-cathode of each SCR in the negative phase-controlled rectifier can be shorted by a transistor switch which is closed when $v_{o,w}$ is positive. A similar inhibit function can be applied to the positive phase-controlled rectifier when $v_{o,w}$ is negative.

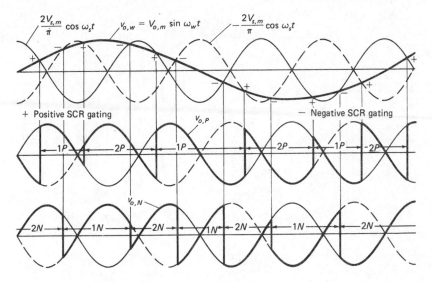

Fig. 8-3. Cosine wave-crossing gating method for Fig. 8-1.

There are a number of additional questions that must be addressed concerning the operation of the single-phase cycloconverter. For example, what are the effects of changes in the magnitude of the load current, the load power factor, and regenerative loads? Reliable operation must be maintained for both continuous and discontinuous load current and for the cases where the load is either receiving power or regenerating to feed power back to the ac supply of the cycloconverter. For the circuits in Fig. 8-1, it is crucial to guarantee that only the N or P thyristors are gated at any given moment. Simultaneous gating of positive and negative thyristors can produce excessive current. To illustrate this fact, assume that both thyristors $1P$ and $1N$ are gated when the transformer secondary voltage is positive on the dot end. This short-circuits the transformer in Fig. 8-1(b), or connects the two secondaries aiding in Fig. 8-1(a) such that an excessively large circulating current flows. In general, when the load power factor is not unity, it is necessary to sense the direction of the load current and then permit gating only that phase-controlled rectifier which can produce load current in the given direction. The phase-control angle is increased to produce negative voltage to drive the load current to zero for the next output-current half-cycle. A short period of zero current is required to assure turn-off of the previously conducting thyristors.

One technique which is used to limit circulating current is to insert inductance in series with the circulating current paths. For the circuit in Fig. 8-1(a), this can be accomplished by the addition of a center-tapped inductor, as shown in Fig. 8-4. With this circuit, the gating signals to the positive- and negative-

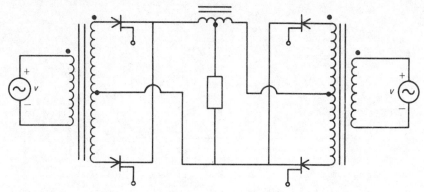

Fig. 8-4. Circulating-current cycloconverter [modification of Fig. 8-1(a)].

controlled rectifiers can be adjusted to produce a controlled circulating current. Generally, the controlled circulating current mode of operation provides smoother control through load current reversals.

8.3. PELLY HARMONIC ANALYSIS [7]

In the previous section, a number of elementary principles of the cycloconverter were described using single-phase circuits. Unfortunately, these circuits have little practical application. Also, even conventional harmonic analysis is quite involved for single-phase cycloconverters.

Pelly [7] presents a generalized harmonic analysis method which can provide quantitative data on the harmonic distortion of three-phase cycloconverter circuits for a wide range of operating conditions. The method can be used to determine the distortion for all frequency components, either above or below the wanted output frequency — not just for strictly harmonic or subharmonic components which are integer multiples or submultiples of the wanted frequency. Figure 8-5 is the circuit which is used to illustrate the harmonic analysis method of Pelly. Since Fig. 8-5 is the modular building block for nearly all practical cycloconverters, this illustrative example forms the basis for the harmonic analysis of practical cycloconverters.

As the first step in this analysis, general expressions are derived for the output-voltage waveforms from each of the phase-controlled rectifiers in Fig. 8-5, referred to as the positive converter and negative converter, assuming that each is in continuous conduction. Both converters are supplied from the same balanced three-phase sinusoidal voltage source.

$$v_{1-n} = V_m \sin \omega_s t \qquad (8\text{-}13)$$

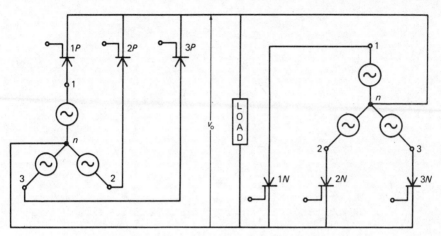

Fig. 8-5. Three-phase, three-pulse cycloconverter.

$$v_{2-n} = V_m \sin(\omega_s t - 120°) \qquad (8\text{-}14)$$

$$v_{3-n} = V_m \sin(\omega_s t + 120°) \qquad (8\text{-}15)$$

Three periodic "switching functions" (see Fig. 8-6) are defined over one power supply period as follows:

$$F_1(\omega_s t - \alpha) = 1; \qquad \frac{\pi}{6} + \alpha < \omega_s t < \frac{5\pi}{6} + \alpha$$

$$= 0; \qquad \text{otherwise} \qquad (8\text{-}16)$$

$$F_2(\omega_s t - \alpha) = 1; \qquad \frac{5\pi}{6} + \alpha < \omega_s t < \frac{3\pi}{2} + \alpha$$

$$= 0; \qquad \text{otherwise} \qquad (8\text{-}17)$$

$$F_3(\omega_s t - \alpha) = 1; \qquad \frac{3\pi}{2} + \alpha < \omega_s t < \frac{13\pi}{6} + \alpha$$

$$= 0; \qquad \text{otherwise} \qquad (8\text{-}18)$$

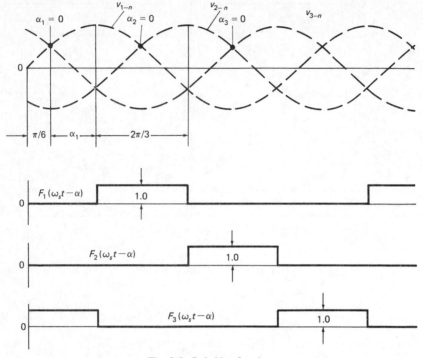

Fig. 8-6. Switching functions.

The Fourier series for $F_1(\omega_s t - \alpha)$ is obtained as follows:

$$A_o = \frac{1}{2\pi} \int_0^{2\pi} F_1(\omega_s t - \alpha)\, d(\omega_s t) = \frac{1}{2\pi} \int_{\alpha+(\pi/6)}^{\alpha+(5\pi/6)} \alpha(\omega_s t) = \frac{1}{3} \qquad (8\text{-}19)$$

With $n \neq 0$,

$$A_n = \frac{1}{\pi} \int_0^{2\pi} F_1(\omega_s t - \alpha) \cos n\omega_s t\, d(\omega_s t) = \frac{1}{\pi} \int_{\alpha+(\pi/6)}^{\alpha+(5\pi/6)} \cos n\omega_s t\, d(\omega_s t)$$

$$= \frac{1}{\pi n} [\sin n\omega_s t]_{\alpha+(\pi/6)}^{\alpha+(5\pi/6)} = \frac{1}{\pi n}\left[\sin n\left(\alpha + \frac{5\pi}{6}\right) - \sin n\left(\alpha + \frac{\pi}{6}\right) \right]$$

$$= \frac{1}{\pi n}\left[\sin n\left(\alpha + \pi - \frac{\pi}{6}\right) - \sin n\left(\alpha + \frac{\pi}{6}\right) \right]$$

$$= \frac{1}{\pi n} \left[\sin n(\alpha + \pi) \cos \frac{n\pi}{6} - \cos n(\alpha + \pi) \sin \frac{n\pi}{6} \right.$$

$$\left. - \sin n\alpha \cos \frac{n\pi}{6} - \cos n\alpha \sin \frac{n\pi}{6} \right]$$

$$= \frac{1}{\pi n} \left[(\sin n\alpha)(\cos n\pi - 1) \cos \frac{n\pi}{6} - (\cos n\alpha)(\cos n\pi + 1) \sin \frac{n\pi}{6} \right]$$

$$(8\text{-}20)$$

or

$$A_n = \frac{-2}{\pi n} \sin n\alpha \cos \frac{n\pi}{6}; \qquad n - \text{odd} \qquad (8\text{-}21)$$

$$A_n = \frac{-2}{\pi n} \cos n\alpha \sin \frac{n\pi}{6}; \qquad n - \text{even} \qquad (8\text{-}22)$$

The B_n coefficients, $(n \neq 0)$, are obtained from

$$B_n = \frac{1}{\pi} \int_0^{2\pi} F_1(\omega_s t - \alpha) \sin n\omega_s t \, d(\omega_s t) = \frac{1}{\pi} \int_{\alpha+(\pi/6)}^{\alpha+(5\pi/6)} \sin n\omega_s t \, d(\omega_s t)$$

$$= \frac{1}{\pi n} [-\cos n\omega_s t]_{\alpha+(\pi/6)}^{\alpha+(5\pi/6)} = \frac{1}{\pi n} \left[\cos n\left(\alpha + \frac{\pi}{6}\right) - \cos n\left(\alpha + \frac{5\pi}{6}\right) \right]$$

$$= \frac{1}{\pi n} \left[\cos n\left(\alpha + \frac{\pi}{6}\right) - \cos n\left(\alpha + \pi - \frac{\pi}{6}\right) \right]$$

$$= \frac{1}{\pi n} \left[\cos n\alpha \cos\left(\frac{n\pi}{6}\right) - \sin n\alpha \sin\left(\frac{n\pi}{6}\right) \right.$$

$$\left. - \cos n(\alpha + \pi) \cos\left(\frac{n\pi}{6}\right) - \sin n(\alpha + \pi) \sin\left(\frac{n\pi}{6}\right) \right]$$

$$= \frac{1}{\pi n} \left[(\cos n\alpha)(1 - \cos n\pi) \cos \frac{n\pi}{6} - (\sin n\alpha)(1 + \cos n\pi) \sin \frac{n\pi}{6} \right]$$

$$(8\text{-}23)$$

or

$$B_n = \frac{2}{\pi n} \cos n\alpha \cos \frac{n\pi}{6}; \quad n - \text{odd} \tag{8-24}$$

$$B_n = \frac{-2}{\pi n} \sin n\alpha \sin \frac{n\pi}{6}; \quad n - \text{even} \tag{8-25}$$

Then, from Eqs. (8-21), (8-22), (8-24), and (8-25), the harmonics of $F_1(\omega_s t - \alpha)$ are

n	A_n	B_n
0	$\dfrac{1}{3}$	0
1	$\dfrac{-\sqrt{3}}{\pi} \sin \alpha$	$\dfrac{\sqrt{3}}{\pi} \cos \alpha$
2	$\dfrac{-\sqrt{3}}{2\pi} \cos 2\alpha$	$\dfrac{-\sqrt{3}}{2\pi} \sin 2\alpha$
3	0	0
4	$\dfrac{-\sqrt{3}}{4\pi} \cos 4\alpha$	$\dfrac{-\sqrt{3}}{4\pi} \sin 4\alpha$
5	$\dfrac{\sqrt{3}}{5\pi} \sin 5\alpha$	$\dfrac{-\sqrt{3}}{5\pi} \cos 5\alpha$
6	0	0
7	$\dfrac{\sqrt{3}}{7\pi} \sin 7\alpha$	$\dfrac{-\sqrt{3}}{7\pi} \cos 7\alpha$
8	$\dfrac{\sqrt{3}}{8\pi} \cos 8\alpha$	$\dfrac{\sqrt{3}}{8\pi} \sin 8\alpha$
9	0	0

10	$\dfrac{\sqrt{3}}{10\pi}\cos 10\alpha$	$\dfrac{\sqrt{3}}{10\pi}\sin 10\alpha$
11	$\dfrac{-\sqrt{3}}{11\pi}\sin 11\alpha$	$\dfrac{\sqrt{3}}{11\pi}\cos 11\alpha$
12	0	0
13	$\dfrac{-\sqrt{3}}{13\pi}\sin 13\alpha$	$\dfrac{\sqrt{3}}{13\pi}\cos 13\alpha$
14	$\dfrac{-\sqrt{3}}{14\pi}\cos 14\alpha$	$\dfrac{-\sqrt{3}}{14\pi}\sin 14\alpha$

.   .   .

.   .   .

.   .   .

Thus, with some trigonometric manipulation, the series representation for $F_1(\omega_s t - \alpha)$ can be written as follows:

$$
\begin{aligned}
F_1(\omega_s t - \alpha) = \frac{1}{3} + \frac{\sqrt{3}}{\pi}\Bigg[&\sin(\omega_s t - \alpha) - \frac{1}{2}\cos 2(\omega_s t - \alpha) \\
&- \frac{1}{4}\cos 4(\omega_s t - \alpha) - \frac{1}{5}\sin 5(\omega_s t - \alpha) \\
&- \frac{1}{7}\sin 7(\omega_s t - \alpha) + \frac{1}{8}\cos 8(\omega_s t - \alpha) \\
&+ \frac{1}{10}\cos 10(\omega_s t - \alpha) + \frac{1}{11}\sin 11(\omega_s t - \alpha) \\
&+ \frac{1}{13}\sin 13(\omega_s t - \alpha) - \frac{1}{14}\cos 14(\omega_s t - \alpha) \\
&+ \dots \Bigg]
\end{aligned}
$$

$$(8\text{-}26)$$

The series representations for $F_2(\omega_s t - \alpha)$ and $F_3(\omega_s t - \alpha)$ are the same as Eq. (8-26), except with $\omega_s t$ replaced by $(\omega_s t - 120°)$ and $(\omega_s t + 120°)$ respectively.

For the continuous-conduction case, zero average output voltage is produced from the positive converter with $\alpha = 90°$. The zero average output voltage waveforms from both the positive and negative converters are shown in Fig. 8-7. Then, to produce an output voltage at the wanted frequency, it is assumed that the gating angles are oscillated to and fro about the $\alpha = 90°$ points. With an arbitrary α-modulation function $\alpha(\omega_w t)$, the general expressions for the positive and negative converter voltages are the following:

$$v_{o,P} = [V_m \sin \omega_s t]\left[F_1\left(\omega_s t - \frac{\pi}{2} + \alpha(\omega_w t)\right)\right]$$

$$+ [V_m \sin(\omega_s t - 120°)]\left[F_2\left(\omega_s t - \frac{\pi}{2} + \alpha(\omega_w t)\right)\right]$$

$$+ [V_m \sin(\omega_s t + 120°)]\left[F_3\left(\omega_s t - \frac{\pi}{2} + \alpha(\omega_w t)\right)\right] \qquad (8\text{-}27)$$

$$v_{o,N} = [V_m \sin \omega_s t]\left[F_1\left(\omega_s t + \frac{\pi}{2} - \alpha(\omega_w t)\right)\right]$$

$$+ [V_m \sin(\omega_s t - 120°)]\left[F_2\left(\omega_s t + \frac{\pi}{2} - \alpha(\omega_w t)\right)\right]$$

$$+ [V_m \sin(\omega_s t + 120°)]\left[F_3\left(\omega_s t + \frac{\pi}{2} - \alpha(\omega_w t)\right)\right] \qquad (8\text{-}28)$$

Substituting the series representations for the switching functions, Eq. (8-26) and the corresponding expressions for F_2 and F_3, the output voltage from the positive converter becomes

$$v_{o,P} = [V_m \sin \omega_s t]\left[\frac{1}{3} + \frac{\sqrt{3}}{\pi}\left\{\sin(\omega_s t - 90° + \alpha(\omega_w t))\right.\right.$$

$$-\frac{1}{2}\cos 2(\omega_s t - 90° + \alpha(\omega_w t))$$

$$\left.\left.-\frac{1}{4}\cos 4(\omega_s t - 90° + \alpha(\omega_w t)) - \ldots\right\}\right]$$

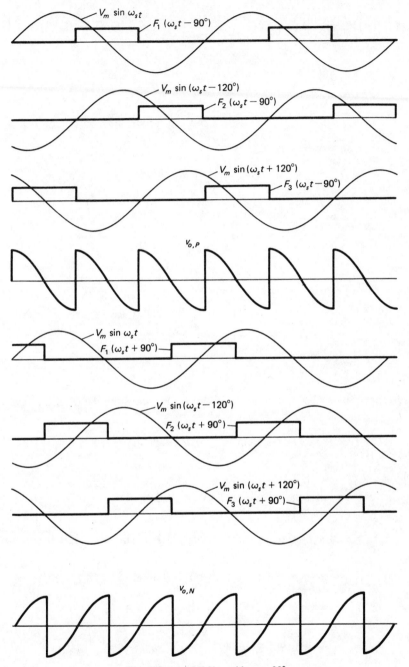

Fig. 8-7. $v_{o,P}$ and $V_{o,N}$ with $\alpha = 90°$.

$$+ [V_m \sin(\omega_s t - 120°)]\left[\frac{1}{3} + \frac{\sqrt{3}}{\pi}\left\{\sin(\omega_s t - 210° + \alpha(\omega_w t))\right.\right.$$

$$- \frac{1}{2}\cos 2(\omega_s t - 210° + \alpha(\omega_w t))$$

$$- \frac{1}{4}\cos 4(\omega_s t - 210° + \alpha(\omega_w t))$$

$$\left.\left.- \cdots \right\}\right]$$

$$+ [V_m \sin(\omega_s t + 120°)]\left[\frac{1}{3} + \frac{\sqrt{3}}{\pi}\left\{\sin(\omega_s t + 30° + \alpha(\omega_w t))\right.\right.$$

$$- \frac{1}{2}\cos 2(\omega_s t + 30° + \alpha(\omega_w t))$$

$$- \frac{1}{4}\cos 4(\omega_s t + 30° + \alpha(\omega_w t))$$

$$\left.\left.- \cdots \right\}\right] \tag{8-29}$$

This equation can be simplified as follows. First, consider the constant term

$$\frac{1}{3}V_m[\sin \omega_s t + \sin(\omega_s t - 120°) + \sin(\omega_s t + 120°)] = 0 \tag{8-30}$$

This term is zero since the sum of three-phase balanced sinusoidal waves is zero. Next, consider the second term of Eq. (8-29):

$$\frac{\sqrt{3}\,V_m}{\pi}\{[\sin \omega_s t][\sin(\omega_s t - 90° + \alpha(\omega_w t))] + [\sin(\omega_s t - 120°)]$$

$$\times [\sin(\omega_s t - 210° + \alpha(\omega_w t))] + [\sin(\omega_s t + 120°)]$$

$$\times [\sin(\omega_s t + 30° + \alpha(\omega_w t))]\}$$

$$= \frac{\sqrt{3}\,V_m}{2\pi}\{\cos[90° - \alpha(\omega_w t)] - \cos[2\omega_s t - 90° + \alpha(\omega_w t)]$$

$$+ \cos[90° - \alpha(\omega_w t)] - \cos[2\omega_s t + 30° + \alpha(\omega_w t)]$$

$$+ \cos[90° - \alpha(\omega_w t)] - \cos[2\omega_s t + 150° + \alpha(\omega_w t)]\}$$

$$= + \frac{3\sqrt{3} V_m}{2\pi} \sin[\alpha(\omega_w t)] \qquad (8\text{-}31)$$

where the identity

$$\cos(A - B) - \cos(A + B) = 2 \sin A \sin B \qquad (8\text{-}32)$$

has been used, and the second harmonic terms add to zero because they are the sum of three equal-amplitude cosine waves displaced 120° from each other.

Now, consider the third term of Eq. (8-29):

$$\frac{-\sqrt{3} V_m}{2\pi} \{[\sin \omega_s t][\cos 2(\omega_s t - 90° + \alpha(\omega_w t))] + [\sin(\omega_s t - 120°)]$$

$$\times [\cos 2(\omega_s t - 210° + \alpha(\omega_w t))] + [\sin(\omega_s t + 120°)]$$

$$\times [\cos 2(\omega_s t + 30° + \alpha(\omega_w t))]\}$$

$$= \frac{-\sqrt{3} V_m}{4\pi} \{\sin(3\omega_s t - 180° + 2\alpha(\omega_w t))$$

$$+ \sin(-\omega_s t + 180° - 2\alpha(\omega_w t))$$

$$+ \sin(3\omega_s t - 180° + 2\alpha(\omega_w t))$$

$$+ \sin(-\omega_s t - 60° - 2\alpha(\omega_w t))$$

$$+ \sin(3\omega_s t + 180° + 2\alpha(\omega_w t))$$

$$+ \sin(-\omega_s t + 60° - 2\alpha(\omega_w t))\}$$

$$= \frac{3\sqrt{3} V_m}{4\pi} \sin[3\omega_s t + 2\alpha(\omega_w t)] \qquad (8\text{-}33)$$

In this case the identity

$$\sin A \cos B = \frac{\sin(A + B) + \sin(A - B)}{2} \qquad (8\text{-}34)$$

is used. Proceeding in a similar fashion for the subsequent terms, Eq. (8-29) becomes

$$v_{o,P} = \frac{3\sqrt{3}\,V_m}{2\pi}\{\sin\,\alpha(\omega_w t)$$

$$+ \frac{1}{2}[\sin 3\omega_s t \cos 2\alpha(\omega_w t) + \cos 3\omega_s t \sin 2\alpha(\omega_w t)]$$

$$+ \frac{1}{4}[\sin 3\omega_s t \cos 4\alpha(\omega_w t) + \cos 3\omega_s t \sin 4\alpha(\omega_w t)]$$

$$+ \frac{1}{5}[\sin 6\omega_s t \cos 5\alpha(\omega_w t) + \cos 6\omega_s t \sin 5\alpha(\omega_w t)]$$

$$+ \frac{1}{7}[\sin 6\omega_s t \cos 7\alpha(\omega_w t) + \cos 6\omega_s t \sin 7\alpha(\omega_w t)] + \ldots\}$$

$$(8\text{-}35)$$

In a similar manner, the output voltage of the negative converter can be written

$$v_{o,N} = \frac{3\sqrt{3}\,V_m}{2\pi}\{\sin\,\alpha(\omega_w t)$$

$$+ \frac{1}{2}[\sin 3\omega_s t \cos 2\alpha(\omega_w t) - \cos 3\omega_s t \sin 2\alpha(\omega_w t)]$$

$$+ \frac{1}{4}[\sin 3\omega_s t \cos 4\alpha(\omega_w t) - \cos 3\omega_s t \sin 4\alpha(\omega_w t)]$$

$$+ \frac{1}{5}[-\sin 6\omega_s t \cos 5\alpha(\omega_w t) + \cos 6\omega_s t \sin 5\alpha(\omega_w t)]$$

$$+ \frac{1}{7}[-\sin 6\omega_s t \cos 7\alpha(\omega_w t) + \cos 6\omega_s t \sin 7\alpha(\omega_w t)] + \ldots\}$$

$$(8\text{-}36)$$

For the circulating-current-mode version of the cycloconverter in Fig. 8-5, the output voltage is

$$v_o = \frac{v_{o,P} + v_{o,N}}{2} \tag{8-37}$$

Using Eqs. (8-35) and (8-36) in Eq. (8-37),

$$v_o = \frac{3\sqrt{3}\,V_m}{2\pi}\left[\sin \alpha(\omega_w t) + \frac{1}{2} \sin 3\omega_s t \cos 2\alpha(\omega_w t) \right.$$

$$+ \frac{1}{4} \sin 3\omega_s t \cos 4\alpha(\omega_w t) + \frac{1}{5} \cos 6\omega_s t \sin 5\alpha(\omega_w t)$$

$$\left. + \frac{1}{7} \cos 6\omega_s t \sin 7\alpha(\omega_w t) + \cdots \right] \tag{8-38}$$

In the circulating-current-free mode of operation, the conduction period of each converter in Fig. 8-5 is determined by the displacement power factor ϕ_w of the load. The positive converter supplies positive load current, and the negative converter supplies the negative half-cycle of output current. Thus, it is necessary to define two additional complementary switching functions designated $F_P(\omega_w t)$ and $F_N(\omega_w t)$. Function $F_P(\omega_w t)$ has unit-amplitude when the positive converter is conducting and zero amplitude otherwise; $F_N(\omega_w t)$ has unit-amplitude when the negative converter is supplying the load current and zero amplitude otherwise. The Fourier series representations of these functions are the following:

$$F_P(\omega_w t) = \frac{1}{2} + \frac{2}{\pi}\left[\sin(\omega_w t + \phi_w) + \frac{1}{3} \sin 3(\omega_w t + \phi_w) \right.$$

$$\left. + \frac{1}{5} \sin 5(\omega_w t + \phi_w) + \frac{1}{7} \sin 7(\omega_w t + \phi_w) + \cdots \right] \tag{8-39}$$

$$F_N(\omega_w t) = \frac{1}{2} - \frac{2}{\pi}\left[\sin(\omega_w t + \phi_w) + \frac{1}{3} \sin 3(\omega_w t + \phi_w) \right.$$

$$\left. + \frac{1}{5} \sin 5(\omega_w t + \phi_w) + \frac{1}{7} \sin 7(\omega_w t + \phi_w) + \cdots \right] \tag{8-40}$$

Then the output voltage of the circulating-current-free cycloconverter is given by

$$v_o = v_{o,P} F_P(\omega_w t) + v_{o,N} F_N(\omega_w t) \tag{8-41}$$

Substituting Eqs. (8-35), (8-36), (8-39), and (8-40) into Eq. (8-41) yields

$$
\begin{aligned}
v_o = {} & \frac{3\sqrt{3}\,V_m}{2\pi} \left\{ \sin\alpha(\omega_w t) + \frac{1}{2}\sin 3\omega_s t \cos 2\alpha(\omega_w t) \right. \\
& + \frac{1}{4}\sin 3\omega_s t \cos 4\alpha(\omega_w t) + \frac{1}{5}\cos 6\omega_s t \sin 5\alpha(\omega_w t) \\
& \left. + \frac{1}{7}\cos 6\omega_s t \sin 7\alpha(\omega_w t) + \ldots \right\} \\
& + \frac{3\sqrt{3}\,V_m}{2\pi} \left\{ \left[\frac{1}{2}\cos 3\omega_s t \sin 2\alpha(\omega_w t) + \frac{1}{4}\cos 3\omega_s t \sin 4\alpha(\omega_w t) \right.\right. \\
& + \frac{1}{5}\sin 6\omega_s t \cos 5\alpha(\omega_w t) \\
& \left. + \frac{1}{7}\sin 6\omega_s t \cos 7\alpha(\omega_w t) + \ldots \right] \\
& \times \frac{4}{\pi} \left[\sin(\omega_w t + \phi_w) + \frac{1}{3}\sin 3(\omega_w t + \phi_w) \right. \\
& \left.\left. + \frac{1}{5}\sin 5(\omega_w t + \phi_w) + \frac{1}{7}\sin 7(\omega_w t + \phi_w) + \ldots \right] \right\}
\end{aligned}
\tag{8-42}
$$

It should be noted that the first part of Eq. (8-42) is identical to Eq. (8-38), the output voltage of the cycloconverter with circulating current.

Finally, the expressions for the output voltage v_o of Fig. 8-5 are determined using the α-modulation function implied by the cosine wave crossing control method. The reference sinusoidal waveform is

$$v_R = rV_{c,m} \sin \omega_w t \tag{8-43}$$

where

$V_{c,m} \triangleq$ peak value of cosine firing waves

$r \triangleq$ a positive constant ≤ 1, which determines the amplitude of the reference wave and thus the wanted output voltage amplitude

Then using the cosine firing waves for the continuous current case, the required gating angles for the positive SCRs (1P, 2P, 3P) on each power supply half cycle are given by

$$V_{c,m} \cos \alpha_{nP} = rV_{c,m} \sin \frac{\omega_w}{\omega_s}\left[(n - 1)\frac{2\pi}{3} + \frac{\pi}{6} + \alpha_{nP}\right] \qquad (8\text{-}11)$$

where

$\alpha_{nP} \triangleq$ gating angle on nth positive half cycle, measured from the $\alpha_{nP} = 0°$ point

$n \triangleq 1, 2, 3, \ldots$ indicates the nth positive half cycle for the three phase supply beginning with v_{1-n} and in the phase sequence $v_{1-n}, v_{2-n}, v_{3-n}, v_{1-n} \ldots$

It should be noted that α_{nP} is the gating angle on the nth positive half cycle measured in supply frequency degrees. Also a given time t_1 measured from the positive going zero crossing of v_{1-n} represents $\omega_w t_1$ wanted frequency degrees, which are related to supply frequency degrees by

$$\omega_w t_1 = \frac{\omega_w}{\omega_s}(\omega_s t_1) \qquad (8\text{-}45)$$

The quiescent or zero-reference input situation is assumed to be a gating angle of 90° producing zero average voltage output from both the positive and negative converters. Then, the to and fro variation in the gating angle from 90° can be written

$$\alpha(\omega_w t) = 90° - \alpha_{nP} \qquad (8\text{-}46)$$

Combining Eqs. (8-44) and (8-46),

$$\alpha(\omega_w t) = \sin^{-1}(r \sin \omega_w t) \tag{8-47}$$

Substituting Eq. (8-47) into Eq. (8-38) yields the cycloconverter output voltage for the circulating current mode:

$$v_o = \frac{3\sqrt{3}\,V_m}{2\pi}\left[r \sin \omega_w t + \frac{1}{2} \sin 3\omega_s t \cos(2 \sin^{-1}(r \sin \omega_w t)) \right.$$

$$+ \frac{1}{4} \sin 3\omega_s t \cos(4 \sin^{-1}(r \sin \omega_w t))$$

$$+ \frac{1}{5} \cos 6\omega_s t \sin(5 \sin^{-1}(r \sin \omega_w t))$$

$$\left. + \frac{1}{7} \cos 6\omega_s t \sin(7 \sin^{-1}(r \sin \omega_w t)) + \ldots \right] \tag{8-48}$$

The first term of Eq. (8-48) is the wanted component of the output voltage. The remaining terms are the harmonic components. If each of these terms can be expressed in terms of simple sinusoidal harmonic functions, then the harmonic components can be determined directly from the resulting expression for v_o.

Consider the last part of the second term in Eq. (8-48),

$$\cos[2 \sin^{-1}(r \sin \omega_w t)] \tag{8-49}$$

This is an even function, and the amplitudes of the cosine terms in the Fourier series for this function are determined as follows:

$$A_n = \frac{1}{\pi} \int_0^{2\pi} \cos[2 \sin^{-1}(r \sin \omega_w t)] \cos n\omega_w t \, d(\omega_w t) \tag{8-50}$$

To simplify the notation, the angular variable $\omega_w t$ is replaced by θ, and then the following substitution is used:

$$x \triangleq \sin^{-1}(r \sin \theta) \tag{8-51}$$

$$\sin x = r \sin \theta \tag{8-52}$$

$$\cos[2 \sin^{-1}(r \sin \theta)] = \cos 2x \tag{8-53}$$

Equation (8-50) is then

$$A_n = \frac{1}{\pi} \int_0^{2\pi} \cos 2x \cos n\theta \, d\theta = \frac{1}{\pi} \int_0^{2\pi} (1 - 2 \sin^2 x) \cos n\theta \, d\theta$$

$$= \frac{1}{\pi} \int_0^{2\pi} (1 - 2r^2 \sin^2 \theta) \cos n\theta \, d\theta \tag{8-54}$$

The first term of the integrand is

$$\frac{1}{\pi} \int_0^{2\pi} \cos n\theta \, d\theta = 0$$

and thus Eq. (8-54) becomes

$$A_n = \frac{-2}{\pi} r^2 \int_0^{2\pi} \sin^2 \theta \cos n\theta \, d\theta = \frac{-r^2}{\pi} \int_0^{2\pi} (1 - \cos 2\theta) \cos n\theta \, d\theta$$

$$= \frac{r^2}{\pi} \int_0^{2\pi} \cos 2\theta \cos n\theta \, d\theta = 0; \qquad n \neq 2 \tag{8-55}$$

For $n = 2$,

$$A_2 = \frac{r^2}{\pi} \int_0^{2\pi} \cos^2 2\theta \, d\theta = \frac{r^2}{\pi} \int_0^{2\pi} \frac{1 + \cos 4\theta}{2} \, d\theta = r^2 \tag{8-56}$$

Thus, the Fourier series expansion for the function (8-49) has only one cosine term — a second harmonic of the wanted frequency — but there is also an average value or dc component.

$$A_o = \frac{1}{2\pi} \int_0^{2\pi} \cos[2 \sin^{-1}(r \sin \theta)] \, d\theta = \frac{1}{2\pi} \int_0^{2\pi} \cos 2x \, d\theta$$

$$= \frac{1}{2\pi} \int_0^{2\pi} (1 - 2 \sin^2 x) \, d\theta = \frac{1}{2\pi} \int_0^{2\pi} (1 - 2r^2 \sin^2 \theta) \, d\theta$$

$$= 1 - \frac{r^2}{\pi} \int_0^{2\pi} \sin^2 \theta \, d\theta = 1 - r^2 \tag{8-57}$$

In summary,

$$\cos 2[\sin^{-1}(r \sin \omega_w t)] = A_{2_0} + A_{2_2} \cos 2\omega_w t \qquad (8\text{-}58)$$

where

$$A_{2_0} \triangleq 1 - r^2; \qquad A_{2_2} \triangleq r^2 \qquad (8\text{-}59)$$

Next, consider the last part of the third term in Eq. (8-48):

$$\cos[4 \sin^{-1}(r \sin \theta)] \qquad (8\text{-}60)$$

Again, this is an even function, and the amplitudes of the cosine terms in the Fourier series are determined from

$$A_o = \frac{1}{2\pi} \int_0^{2\pi} \cos[4 \sin^{-1}(r \sin \theta)] d\theta \qquad (8\text{-}61)$$

and

$$A_n = \frac{1}{\pi} \int_0^{2\pi} \cos[4 \sin^{-1}(r \sin \theta)] \cos n\theta \, d\theta \qquad (8\text{-}62)$$

Using the substitution of Eq. (8-51), Eq. (8-61) becomes

$$A_o = \frac{1}{2\pi} \int_0^{2\pi} \cos 4x \, d\theta = \frac{1}{2\pi} \int_0^{2\pi} (1 - 2 \sin^2 2x) \, d\theta$$

$$= \frac{1}{2\pi} \int_0^{2\pi} [1 - 2(2 \sin x \cos x)^2] d\theta = \frac{1}{2\pi} \int_0^{2\pi} [1 - 8 \sin^2 x (1 - \sin^2 x)] d\theta$$

$$= \frac{1}{2\pi} \int_0^{2\pi} (1 - 8r^2 \sin^2\theta + 8r^4 \sin^4\theta) \, d\theta$$

$$= 1 - \frac{4r^2}{\pi} \int_0^{2\pi} \sin^2\theta \, d\theta + \frac{4r^4}{\pi} \int_0^{2\pi} \sin^4\theta \, d\theta$$

$$= 1 - 4r^2 + \frac{4r^4}{\pi} \left[\frac{3\theta}{8} - \frac{\sin 2\theta}{4} + \frac{\sin 4\theta}{32} \right] \Big|_0^{2\pi} = 1 - 4r^2 + 3r^4 \qquad (8\text{-}63)$$

From Eqs. (8-51) and (8-62),

$$A_n = \frac{1}{\pi} \int_0^{2\pi} \cos 4x \, \cos n\theta \, d\theta = \frac{1}{\pi} \int_0^{2\pi} (1 - 8 \sin^2 x + 8 \sin^4 x) \cos n\theta \, d\theta$$

$$= \frac{1}{\pi} \int_0^{2\pi} (1 - 8r^2 \sin^2\theta + 8r^4 \sin^4\theta) \cos n\theta \, d\theta$$

$$= \frac{-8r^2}{\pi} \int_0^{2\pi} \sin^2\theta \, \cos n\theta \, d\theta + \frac{8r^4}{\pi} \int_0^{2\pi} \sin^4\theta \, \cos n\theta \, d\theta$$

$$= \frac{-4r^2}{\pi} \int_0^{2\pi} (1 - \cos 2\theta) \cos n\theta \, d\theta + \frac{2r^4}{\pi} \int_0^{2\pi} (1 - \cos 2\theta)^2 \cos n\theta \, d\theta$$

$$= \frac{4r^2}{\pi} \int_0^{2\pi} \cos 2\theta \, \cos n\theta \, d\theta + \frac{2r^4}{\pi} \int_0^{2\pi} (1 - 2 \cos 2\theta + \cos^2 2\theta) \cos n\theta \, d\theta$$

$$= \frac{4r^2}{\pi} \left[\int_0^{2\pi} \cos 2\theta \, \cos n\theta \, d\theta \right.$$

$$\left. + \frac{r^2}{4} \int_0^{2\pi} (-4 \cos 2\theta + 3 + \cos 4\theta) \cos n\theta \, d\theta \right] \qquad (8\text{-}64)$$

This equation implies that $A_n = 0$, except when $n = 2$ and $n = 4$.

$$A_2 = \frac{4r^2}{\pi} \left[\int_0^{2\pi} \cos^2 2\theta \, d\theta - r^2 \int_0^{2\pi} \cos^2 2\theta \, d\theta \right] = 4r^2(1 - r^2) \qquad (8\text{-}65)$$

$$A_4 = \frac{r^4}{\pi} \int_0^{2\pi} \cos^2 4\theta \, d\theta = r^4 \qquad (8\text{-}66)$$

Therefore,

$$\cos[4 \sin^{-1}(r \sin \omega_w t)] = A_{4_o} + A_{4_2} \cos 2\omega_w t + A_{4_4} \cos 4\omega_w t \qquad (8\text{-}67)$$

where

$$A_{4_o} \triangleq 1 - 4r^2 + 3r^4; \qquad A_{4_2} \triangleq 4r^2(1 - r^2); \qquad A_{4_4} \triangleq r^4 \qquad (8\text{-}68)$$

Next, consider the last part of the fourth term in Eq. (8-48):

$$\sin[5 \sin^{-1}(r \sin \theta)] \tag{8-69}$$

This is an odd function, and the amplitudes of the sine terms in the Fourier series are determined from

$$B_n = \frac{1}{\pi} \int_0^{2\pi} \sin[5 \sin^{-1}(r \sin \theta)] \sin n\theta \, d\theta \tag{8-70}$$

Again using Eq. (8-51),

$$B_n = \frac{1}{\pi} \int_0^{2\pi} \sin 5x \sin n\theta \, d\theta$$

$$= \frac{1}{\pi} \int_0^{2\pi} [5 \sin x - 20 \sin^3 x + 16 \sin^5 x] \sin n\theta \, d\theta$$

$$= \frac{1}{\pi} \int_0^{2\pi} [5r \sin \theta - 20r^3 \sin^3 \theta + 16r^5 \sin^5 \theta] \sin n\theta \, d\theta$$

$$= \frac{1}{\pi} \int_0^{2\pi} \left[5r \sin \theta - 20r^3 \frac{(3 \sin \theta - \sin 3\theta)}{4} \right.$$

$$\left. + 16r^5 \frac{(\sin 5\theta - 5 \sin \theta + 20 \sin^3 \theta)}{16} \right] \sin n\theta \, d\theta$$

$$= \frac{1}{\pi} \int_0^{2\pi} [5r \sin \theta - 15r^3 \sin \theta + 5r^3 \sin 3\theta + r^5 \sin 5\theta - 5r^5 \sin \theta$$

$$+ 5r^5(3 \sin \theta - \sin 3\theta)] \sin n\theta \, d\theta$$

$$= \frac{1}{\pi} \int_0^{2\pi} [5r(1 - 3r^2 + 2r^4) \sin \theta + 5r^3(1 - r^2) \sin 3\theta$$

$$+ r^5 \sin 5\theta] \sin n\theta \, d\theta \tag{8-71}$$

Thus, in this case only the $n = 1$, 3, and 5 terms are nonzero, and

$$B_1 = \frac{5r(1 - 3r^2 + 2r^4)}{\pi} \int_0^{2\pi} \sin^2\theta \, d\theta = 5r(1 - 3r^2 + 2r^4) \tag{8-72}$$

$$B_3 = \frac{5r^3(1 - r^2)}{\pi} \int_0^{2\pi} \sin^2 3\theta \, d\theta = 5r^3(1 - r^2) \qquad (8\text{-}73)$$

$$B_5 = \frac{r^5}{\pi} \int_0^{2\pi} \sin^2 5\theta \, d\theta = r^5 \qquad (8\text{-}74)$$

Therefore,

$$\sin[5 \sin^{-1}(r \sin \omega_w t)] = B_{5_1} \sin \omega_w t + B_{5_3} \sin 3\omega_w t + B_{5_5} \sin 5\omega_w t \qquad (8\text{-}75)$$

where

$$B_{5_1} \triangleq 5r(1 - 3r^2 + 2r^4); \qquad B_{5_3} \triangleq 5r^3(1 - r^2); \qquad B_{5_5} \triangleq r^5 \qquad (8\text{-}76)$$

Finally, the last part of the fifth term in Eq. (8-48) is

$$\sin[7 \sin^{-1}(r \sin \theta)] \qquad (8\text{-}77)$$

This is again an odd function, and the amplitudes of the sine terms in its Fourier series are

$$B_n = \frac{1}{\pi} \int_0^{2\pi} \sin[7 \sin^{-1}(r \sin \theta)] \sin n\theta \, d\theta \qquad (8\text{-}78)$$

Using Eq. (8-51), this becomes

$$B_n = \frac{1}{\pi} \int_0^{2\pi} \sin 7x \sin n\theta \, d\theta$$

$$= \frac{1}{\pi} \int_0^{2\pi} [7 - 56 \sin^2 x + 112 \sin^4 x - 64 \sin^6 x] \sin x \sin n\theta \, d\theta$$

$$= \frac{r}{\pi} \int_0^{2\pi} (7 \sin \theta - 56r^2 \sin^3 \theta + 112r^4 \sin^5 \theta - 64r^6 \sin^7 \theta) \sin n\theta \, d\theta \qquad (8\text{-}79)$$

The following trigonometric relationships are useful:

$$\sin^3 \theta = \frac{3 \sin \theta - \sin 3\theta}{4} \qquad (8\text{-}80)$$

$$\sin^5\theta = \frac{10 \sin \theta - 5 \sin 3\theta + \sin 5\theta}{16} \tag{8-81}$$

$$\sin^7\theta = \frac{35 \sin \theta - 21 \sin 3\theta + 7 \sin 5\theta - \sin 7\theta}{64} \tag{8-82}$$

Substituting Eqs. (8-80) through (8-82) in Eq. (8-79) yields

$$B_n = \frac{r}{\pi} \int_0^{2\pi} [7 \sin \theta - 14r^2(3 \sin \theta - \sin 3\theta)$$

$$+ 7r^4(10 \sin \theta - 5 \sin 3\theta + \sin 5\theta)$$

$$- r^6(35 \sin \theta - 21 \sin 3\theta + 7 \sin 5\theta - \sin 7\theta)] \sin n\theta \, d\theta$$

$$= \frac{r}{\pi} \int_0^{2\pi} [(7 - 42r^2 + 70r^4 - 35r^6) \sin \theta$$

$$+ (14r^2 - 35r^4 + 21r^6) \sin 3\theta + (7r^4 - 7r^6) \sin 5\theta$$

$$+ r^6 \sin 7\theta] \sin n\theta \, d\theta \tag{8-83}$$

In this instance, only the $n = 1, 3, 5,$ and 7 terms are nonzero, and

$$\sin[7 \sin^{-1}(r \sin \omega_w t)] = B_{7_1} \sin \omega_w t + B_{7_3} \sin 3\omega_w t + B_{7_5} \sin 5\omega_w t$$

$$+ B_{7_7} \sin 7\omega_w t \tag{8-84}$$

where

$$\left.\begin{aligned}
B_{7_1} &\triangleq 7r(1 - 6r^2 + 10r^4 - 5r^6) \\
B_{7_3} &\triangleq 7r^3(2 - 5r^2 + 3r^4) \\
B_{7_5} &\triangleq 7r^5(1 - r^2) \\
B_{7_7} &\triangleq r^7
\end{aligned}\right\} \tag{8-85}$$

Using the results given in Eqs. (8-58), (8-59), (8-67), (8-68), (8-84), and (8-85), the output voltage expression (8-48) for the circulating-current-mode cycloconverter becomes

$$v_o = \frac{3\sqrt{3}\,V_m}{2\pi}\left\{ r\,\sin\omega_w t + \left(\frac{1}{2}\sin 3\omega_s t\right)\left[1 - r^2 + r^2\cos 2\omega_w t\right] \right.$$

$$+ \left(\frac{1}{4}\sin 3\omega_s t\right)\left[1 - 4r^2 + 3r^4 + 4r^2(1 - r^2)\cos 2\omega_w t\right.$$

$$\left. + r^4\cos 4\omega_w t\right]$$

$$+ \left(\frac{1}{5}\cos 6\omega_s t\right)\left[5r(1 - 3r^2 + 2r^4)\sin\omega_w t\right.$$

$$\left. + 5r^3(1 - r^2)\sin 3\omega_w t + r^5\sin 5\omega_w t\right]$$

$$+ \left(\frac{1}{7}\cos 6\omega_s t\right)\left[7r(1 - 6r^2 + 10r^4 - 5r^6)\sin\omega_w t\right.$$

$$+ 7r^3(2 - 5r^2 + 3r^4)\sin 3\omega_w t$$

$$+ 7r^5(1 - r^2)\sin 5\omega_w t$$

$$\left. + r^7\sin 7\omega_w t\right] + \dots\right\} \tag{8-86}$$

This expression can be rearranged as follows:

$$v_o = \frac{3\sqrt{3}\,V_m}{2\pi}\left\{ r\,\sin\omega_w t + \left[\frac{1}{2}(1 - r^2) + \frac{1}{4}(1 - 4r^2 + 3r^4)\right]\sin 3\omega_s t \right.$$

$$+ \left[\frac{1}{2}r^2 + r^2(1 - r^2)\right]\sin 3\omega_s t\,\cos 2\omega_w t$$

$$+ \frac{r^4}{4}\sin 3\omega_s t\,\cos 4\omega_w t$$

$$+ \left[r(1 - 3r^2 + 2r^4) + r(1 - 6r^2 + 10r^4 - 5r^6)\right]$$

$$\times \cos 6\omega_s t\,\sin\omega_w t$$

$$+ \left[r^3(1 - r^2) + r^3(2 - 5r^2 + 3r^4)\right]\cos 6\omega_s t\,\sin 3\omega_w t$$

$$+ \left[\frac{r^5}{5} + r^5(1 - r^2)\right]\cos 6\omega_s t\,\sin 5\omega_w t$$

$$+ \frac{r^7}{7} \cos 6\omega_s t \, \sin 7\omega_w t + \dots \Big\}$$

$$= \frac{3\sqrt{3}\,V_m}{2\pi} \Big\{ r \sin \omega_w t + \frac{3}{4}(1 - 2r^2 + r^4) \sin 3\omega_s t$$

$$+ \frac{r^2}{4}(3 - 2r^2)[\sin(3\omega_s t + 2\omega_w t) + \sin(3\omega_s t - 2\omega_w t)]$$

$$+ \frac{r^4}{8}[\sin(3\omega_s t + 4\omega_w t) + \sin(3\omega_s t - 4\omega_w t)]$$

$$+ \frac{r}{2}(2 - 9r^2 + 12r^4 - 5r^6)$$

$$\times [\sin(6\omega_s t + \omega_w t) - \sin(6\omega_s t - \omega_w t)]$$

$$+ \frac{3r^3}{2}(1 - 2r^2 + r^4)[\sin(6\omega_s t + 3\omega_w t) - \sin(6\omega_s t - 3\omega_w t)]$$

$$+ \frac{r^5}{10}(6 - 5r^2)[\sin(6\omega_s t + 5\omega_w t) - \sin(6\omega_s t - 5\omega_w t)]$$

$$+ \frac{r^7}{14}[\sin(6\omega_s t + 7\omega_w t) - \sin(6\omega_s t - 7\omega_w t)] + \dots \Big\}$$

$$(8\text{-}87)$$

With $r = 1$, Eq. (8-87) simplifies to

$$v_o = \frac{3\sqrt{3}\,V_m}{2\pi} \Big[\sin \omega_w t + \frac{1}{4} \sin(3\omega_s t + 2\omega_w t) + \frac{1}{4} \sin(3\omega_s t - 2\omega_w t)$$

$$+ \frac{1}{8} \sin(3\omega_s t + 4\omega_w t) + \frac{1}{8} \sin(3\omega_s t - 4\omega_w t)$$

$$+ \frac{1}{10} \sin(6\omega_s t + 5\omega_w t) - \frac{1}{10} \sin(6\omega_s t - 5\omega_w t)$$

$$+ \frac{1}{14} \sin(6\omega_s t + 7\omega_w t) - \frac{1}{14} \sin(6\omega_s t - 7\omega_w t) + \dots \Big]$$

$$(8\text{-}88)$$

This is the form of v_o desired since all frequency components in the output waveform are evident. However, Eq. (8-88) has been obtained only after an extensive excursion into the realm of trigonometric identity manipulations. It should be noted that only the first terms of Eq. (8-88) have been derived, and this equation has an infinite number of terms. However, the amplitudes of subsequent terms become progressively smaller, so they are not of great practical significance.

Pelly [7] includes extensive tables for use in the determination of the terms in Eq. (8-88) as well as some additional terms. Also, a similar development is carried out for the circulating-current-free cycloconverter. This latter development is more involved than the derivation of Eq. (8-88) because there are additional terms in the v_o expression which result in infinite series expressions instead of only the finite series expansions as in Eqs. (8-58), (8-67), (8-75), and (8-84).

The object of this discussion has been to illustrate the application of the powerful harmonic analysis technique described by Pelly. Although it involves extensive trigonometric manipulation, the resulting expressions indicate clearly the wanted and unwanted frequency components in cycloconverter output-voltage waveforms. A serious study of Pelly [1] is recommended for one interested in polyphase cycloconverter analysis and design.

REFERENCES

1. Pelly, B. R. *Thyristor Phase-Controlled Converters and Cycloconverters*. New York: John Wiley & Sons, 1971.
2. McMurray, William. *The Theory and Design of Cycloconverters*. Cambridge, MA: MIT Press, 1972.
3. Dewan, S. B., and Straughen, A. *Power Semiconductor Circuits*. New York: John Wiley & Sons, 1975, pp. 468–491.
4. Gyugyi, L., and Pelly, B. R. *Static Power Frequency Changers*. New York: John Wiley & Sons, 1976.
5. Pelly, B. R. Op. cit., pp. 229–238.
6. Dewan, S. B., and Straughen, A. Op. cit., pp. 474–489.
7. Pelly, B. R. Op. cit., chap. 11.

PROBLEMS

1. Derive the Fourier series representation shown in Eq. (8-10) for the waveform of Fig. 8-2(d).
2. Derive the Fourier series representation for the waveform in Fig. 8-2(e), where the gating angle α is 90° on the first, third, fourth, and sixth half-cycles.
3. For the single-phase cycloconverter of Fig. 8-1(a) with a resistive load, use the cosine wave crossing (CWC) gating principle to determine the gating

angle sequence for operation with a waveform of the type in Fig. 8-2(e) when the reference sinusoidal wave has the same amplitude as the cosine gating waves. Then calculate the harmonics as a percentage of the fundamental and compare these results with those for the waveform of Problem 2 as given in the text.

4. Consider the cycloconverter of Fig. 8-5 assuming that the cosine wave crossing gating method is used with the gating waves for the continuous current case. Calculate the gating angles $(\alpha_{1P}, \alpha_{2P}, \alpha_{3P})$ on the first three positive half cycles of the supply frequency with $v_{1-n} = 300 \sin 2513t$ and a reference sine wave $v_R = 200 \sin 377t$.

5. Write the expression equivalent to (8-44) but for the negative SCRs (1N, 2N, 3N) in the circuit of Fig. 8-5, and then repeat problem 4 for the negative SCRs.

6. Explain in detail why the $v_{o,N}$ waveform is as shown in Fig. 8-7.

7. Derive all of the terms in Eq. (8-35) from Eq. (8-29).

8. Show that Eq. (8-36) is correct.

9. Derive Eqs. (8-39) and (8-40).

10. Prove that Eq. (8-49) is an even function.

11. Prove that Eq. (8-69) is an odd function.

9
ac PHASE CONTROL

9.1. SINGLE PHASE

9.1.1. Resistive Load

One of the simplest single-phase ac phase-control circuits is shown in Fig. 9-1. The voltage impressed on the load is controlled by changing the phase-control angle α. Usually a symmetrical output is required so that each SCR is gated at the same α. Of course, a given SCR is gated only on alternate half-cycles. It is possible theoretically to use a single triac in place of the inverse-parallel pair of SCRs. However, the dv/dt capability and other rating limitations of presently available triacs often make necessary the use of SCRs. In this chapter, the inverse-parallel pair of SCRs will always be considered since this arrangement is functionally equivalent to the triac and is more generally applicable.

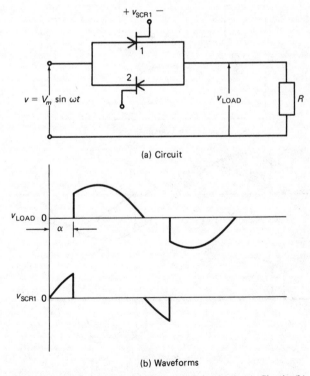

(a) Circuit

(b) Waveforms

Fig. 9-1. Single-phase ac phase-control circuit with resistive load. (a) Circuit; (b) waveforms.

203

9.1.2. R-L Load

Figure 9-2 illustrates this circuit. The waveforms of Fig. 9-2(b) show the operation when the current is discontinuous. When α is advanced to a value equal to the load power factor angle ϕ, where

$$\phi = \tan^{-1}\frac{\omega L}{R} \tag{9-1}$$

a change in the mode of operation occurs. With α less than ϕ, there is 180° conduction in each thyristor. Over the range of firing angles $0 < \alpha < \phi$, the operation will remain the same since the next SCR to conduct does not conduct at the first instant of gating, but rather when the circuit current reverses. Thus,

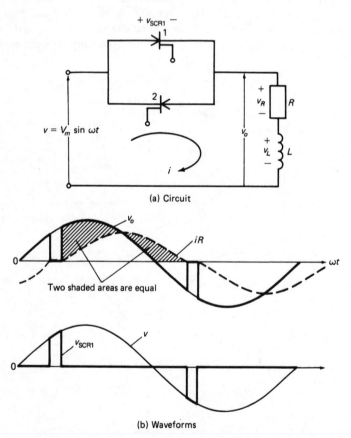

(a) Circuit

(b) Waveforms

Fig. 9-2. Single-phase ac phase-control circuit with *R-L* load. (a) circuit; (b) waveforms.

the load current and voltage waveforms remain the same over the range of gating angles from zero to the power factor angle ϕ.

It is important to note that, in general, pulse gating cannot be used for the single-phase ac circuit with R-L load. For example, consider the following situation. Assume that the circuit of Fig. 9-2 is off; that is, no gating signals are being supplied. Then, suppose that very short duration gating signals are initiated with α less than the circuit power factor angle. The first SCR to be gated is the only one that will ever conduct. After the ac voltage reverses, the current does not reverse until some time after the voltage reversal, as determined by the circuit power factor. If the short duration gating pulse for the next SCR ends before the circuit current has reversed, the previously conducting SCR will remain on. It will go off when the circuit current attempts to reverse. Since the gating signal for the other SCR is no longer present, neither SCR will conduct until the SCR which conducted first is gated again. Thus, only one SCR will conduct, its conduction will occur on alternate half-cycles, and unidirectional current will flow through the load.

9.2. THREE PHASE

9.2.1. Resistive Load

One basic three-phase ac voltage control circuit is shown in Fig. 9-3. This scheme includes an inverse-parallel pair of SCRs in series with each ac line and a wye-connected balanced resistive load. The supply is assumed to be a three-phase balanced sinusoidal voltage source as follows:

$$v_{1-2} = V_m \sin \omega t \tag{9-2}$$

$$v_{2-3} = V_m \sin(\omega t - 120°) \tag{9-3}$$

$$v_{3-1} = V_m \sin(\omega t + 120°) \tag{9-4}$$

With this definition of the ac line-to-line voltages, the SCR numbering system of Fig. 9-3 again produces the situation (as for the three-phase bridge-controlled rectifier) where the next SCR to conduct is always the device with the next higher number.

Before discussing the detailed waveforms for the circuit of Fig. 9-3, two important features of this ac phase control must be noted. First, the $\alpha = 0°$ point is the intersection point of two consecutive ac line-to-line voltages. This is not the same point as for the three-phase bridge-controlled rectifier. In that case, the $\alpha = 0°$ point was the intersection of a positive line-to-line voltage and the next negative line-to-line voltage.

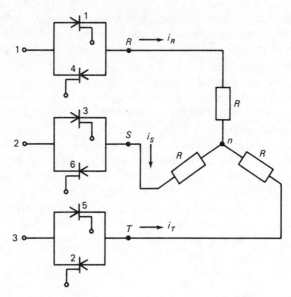

Fig. 9-3. Three-phase wye-connected ac voltage control with resistive load.

Second, there are three possible modes of operation for the circuit in Fig. 9-3.

Mode OFF. All devices are off.
   Mode II. One device conducts in each of only two ac lines; two devices total are conducting.
   Mode III. One device conducts in all three ac lines, i.e., three devices are conducting simultaneously.

The duration of the gating signals is also important for the circuit similar to Fig. 9-3 when the load is not purely resistive. Thus, for the following discussion, 165° long gating signals are assumed.

Now, the following step-by-step procedure is used to determine the waveforms for Fig. 9-3 with a given α. (Note: It is assumed that the gating signals for the given α are plotted on three-phase paper, as shown in Fig. 9-4.)

Preparation.   Starting at the instant when the gating signal to SCR1 is initiated, assume that all SCRs are conducting which have a gating signal.

Step 1 (subsequent intervals).   If the previous interval ended when a new SCR was gated, assume that it is conducting as well as all other SCRs which have a gating signal. If the previous interval ended when an SCR forward current stopped, assume this particular SCR is off but that all other SCRs previously conducting continue to conduct.

Step 2. Check that the current flowing in each ac line is a forward current for the SCR assumed to be conducting.

a) If an SCR is conducting in each ac line (Mode III), the load voltages are

$$v_{R-n} = v_{1-N} \tag{9-5}$$

$$v_{S-n} = v_{2-N} \tag{9-6}$$

$$v_{T-n} = v_{3-N} \tag{9-7}$$

(Note: The symbol n designates the neutral of the balanced three-phase resistive load, and the symbol N designates the neutral of the balanced three-phase voltage source.) The direction of the current in each ac line is determined by the polarities of v_{1-N}, v_{2-N}, and v_{3-N} which determine the polarities of v_{R-n}, v_{S-n}, and v_{T-n}, respectively. The polarity of the voltage drop across a given resistor R defines the direction of current flow.

b) If Mode II operation occurs, the current in one line is zero and the direction of current in the other two lines is determined by the polarity of the particular line-to-line voltage.

Step 3. Determine when the end of the particular interval occurs, which is when one or more of the conducting SCRs becomes reverse-biased or when a new SCR is gated.

As an illustration of the use of the procedure, waveforms for v_{R-S} and v_{SCR1} are plotted in Fig. 9-4(a) for $\alpha = 15°$.

Interval A. $(45° - ?)$

Step 1: SCR1, 5, and 6 have gating signals starting at $\omega t = 45°$, so they are assumed to conduct—Mode III$_{1,5-6}$.

(Note: In general, the mode subscript indicates the one or two odd-numbered SCRs conducting positive line current, followed by the one or two even-numbered SCRs conducting negative line current.)

Step 2: $v_{R-n} = v_{1-N} > 0$, so forward current in SCR1.
$v_{S-n} = v_{2-N} < 0$, so forward current in SCR6.
$v_{T-n} = v_{3-N} > 0$, so forward current in SCR5.

($v_{R-S} = v_{1-2}$ and $v_{SCR1} = 0$.)

Step 3: $v_{T-n} = v_{3-N} < 0$ after $\omega t = 90°$, so forward current in SCR5 ceases.

(Interval A ends at $\omega t = 90°$.)

Interval B. $(90° - ?)$

Step 1: SCR5 is assumed to be off, and SCR1 and 6 are assumed to be on since they were previously conducting — Mode II_{1-6}.

Step 2: $v_{R-S} = v_{1-2} > 0$, so forward current in both SCR1 and SCR6.

$(v_{R-S} = v_{1-2}$ and $v_{SCR1} = 0.)$

Step 3: SCR2 is gated at $\omega t = 105°$, so this is the end of this interval.

(Interval B ends at $\omega t = 105°$.)

Interval C. $(105° - ?)$

Step 1: SCR1, 2, and 6 are assumed to conduct — Mode $III_{1-2,6}$.

Step 2: $v_{R-n} = v_{1-N} > 0$, so forward current in SCR1.

$v_{S-n} = v_{2-N} < 0$, so forward current in SCR6.

$v_{T-n} = v_{3-N} < 0$, so forward current in SCR2.

$(v_{R-S} = v_{1-2}$ and $v_{SCR1} = 0.)$

Step 3: $v_{S-n} = v_{2-N} > 0$ after $\omega t = 150°$, so forward current in SCR6 ceases.

(Interval C ends at $\omega t = 150°$.)

Interval D. $(150° - ?)$

Step 1: SCR6 is assumed off and SCR1 and 2 are assumed to continue to conduct — Mode II_{1-2}.

Step 2: $v_{R-T} = v_{1-3} > 0$, so forward current in both SCR1 and SCR2.

$(v_{R-S} = v_{1-3}/2$ and $v_{SCR1} = 0.)$

Step 3: SCR3 is gated at $165°$, so this ends the interval.

(Interval D ends at $165°$.)

Interval E. $(165° - ?)$

Step 1: SCR1, 2, and 3 are assumed conducting — Mode $III_{1,3-2}$.

Step 2: $v_{R-n} = v_{1-N} > 0$, so forward current in SCR1.

$v_{S-n} = v_{2-N} > 0$, so forward current in SCR3.

$v_{T-n} = v_{3-N} < 0$, so forward current in SCR2.

$(v_{R-S} = v_{1-2}$, and $v_{SCR1} = 0.)$

Step 3: $v_{R-n} = v_{1-N} < 0$ after $\omega t = 210°$, so forward current in SCR1 ceases.

(Interval E ends at 210°.)

Interval F.   (210° – ?)

Step 1: SCR1 is assumed off and SCR2 and 3 are assumed to continue to conduct — Mode II_{3-2}.
Step 2: $v_{S-T} = v_{2-3} > 0$, so forward current in both SCR2 and 3.

($v_{R-S} = -v_{2-3}/2$ and $v_{SCR1} = v_{1-2} + v_{2-3}/2$.)

Step 3: SCR4 is gated at 225°, which ends this interval.

(Interval F ends at 225°.)

Interval G.   (225° – ?)

Step 1: SCR2, 3, and 4 are assumed conducting — Mode $III_{3-2,4}$.
Step 2: $v_{R-n} = v_{1-N} < 0$, so forward current in SCR4.
$v_{S-n} = v_{2-N} > 0$, so forward current in SCR3.
$v_{T-n} = v_{3-N} < 0$, so forward current in SCR2.

($v_{R-S} = v_{1-2}$ and $v_{SCR1} = 0$.)

Step 3: v_{3-N} reverses polarity at 270°, so forward current in SCR2 ceases.

(Interval G ends at 270°.)

Interval H.   (270° – ?)

Step 1: SCR3 and 4 are assumed to conduct — Mode II_{3-4}.
Step 2: $v_{R-S} = v_{1-2} < 0$, so forward current in SCR3 and 4.

($v_{R-S} = v_{1-2}$ and $v_{SCR1} = 0$.)

Step 3: SCR5 is gated at 285°, which ends this interval.

(Interval H ends at 285°.)

Interval I.   (285° – ?)

Step 1: SCR3, 4, and 5 are assumed conducting — Mode $III_{3,5-4}$.

Step 2: $v_{R-n} = v_{1-N} < 0$, so forward current in SCR4.
$v_{S-n} = v_{2-N} > 0$, so forward current in SCR3.
$v_{T-n} = v_{3-N} > 0$, so forward current in SCR5.

($v_{R-S} = v_{1-2}$ and $v_{SCR1} = 0$.)

Step 3: v_{2-N} becomes negative at 330° so SCR3 stops conducting.

(Interval I ends at 330°.)

Interval J. (330° − ?)

Step 1: SCR4 and 5 are assumed conducting—Mode II$_{5-4}$.
Step 2: $v_{T-R} = v_{3-1} > 0$, so forward current in SCR4 and 5.

($v_{R-S} = v_{1-3}/2$ and $v_{SCR1} = 0°$.)

Step 3: SCR6 is gated at 345°, which ends this interval.

(Interval J ends at 345°.)

Interval K. (345° − ?)

Step 1: SCR4, 5, and 6 are assumed conducting—Mode III$_{5-4,6}$.
Step 2: $v_{R-n} = v_{1-N} < 0$, so forward current in SCR4.
$v_{S-n} = v_{2-N} < 0$, so forward current in SCR6.
$v_{T-n} = v_{3-N} > 0$, so forward current in SCR5.

($v_{R-S} = v_{1-2}$ and $v_{SCR1} = 0$.)

Step 3: v_{1-N} becomes positive at 390°, so SCR4 stops conducting.

(Interval K ends at 390°.)

Interval L. (390° − ?)

Step 1: SCR5 and 6 are assumed to conduct—Mode II$_{5-6}$.
Step 2: $v_{S-T} = v_{2-3} < 0$, so forward current in SCR5 and 6.

($v_{R-S} = v_{3-2}/2$ and $v_{SCR1} = v_{1-2} - v_{3-2}/2$.)

Step 3: SCR1 is gated at 405°, which ends this interval.

(Interval L ends at 405°.)

Interval M.   (405° − ?)

This is the same as Interval A, so now waveform repeats.

Figure 9-4(a) shows the waveforms of v_{R-S} and v_{SCR1} for all intervals of operation with $\alpha = 15°$. Using the same procedure, waveforms can be obtained for any other gating angle. Figure 9-4(b) through 9-4(e) show the waveforms for $\alpha = 45°$, 75°, 105°, and 135°, respectively.

9.2.2. *R-L* Load

As illustrated in the previous section, it is quite involved to sketch the waveforms for the circuit of Fig. 9-3, even with resistive load. It becomes even more involved when there is an *R-L* load. In fact, to attempt to sketch waveforms for Fig. 9-3 is not very practical except for the resistive-load case. With

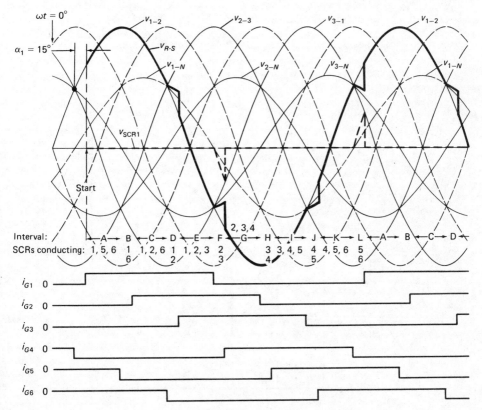

Fig. 9-4(a). Waveforms for Fig. 9-3 with $\alpha = 15°$.

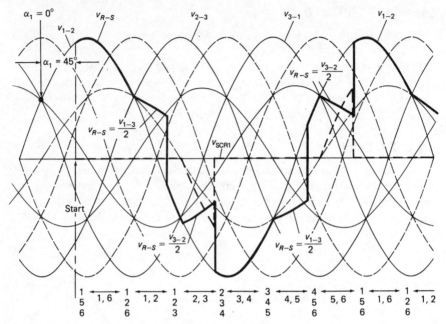

Fig. 9-4(b). Waveforms for Fig. 9-3 with $\alpha = 45°$.

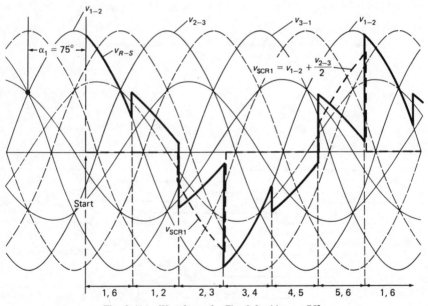

Fig. 9-4(c). Waveforms for Fig. 9-3 with $\alpha = 75°$.

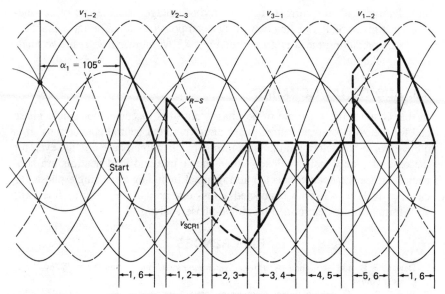

Fig. 9-4(d). Waveforms for Fig. 9-3 with $\alpha = 105°$.

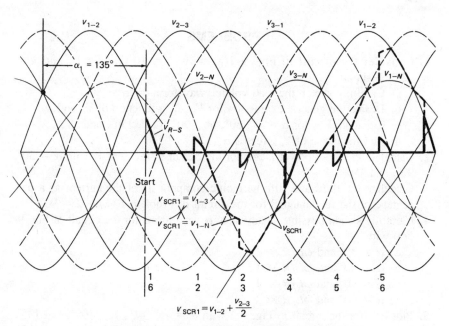

$$v_{SCR1} = v_{1-2} + \frac{v_{2-3}}{2}$$

Fig. 9-4(e). Waveforms for Fig. 9-3 with $\alpha = 135°$.

an inductive load, the current lags the driving voltage, and the end of conduction for a given SCR is a function of the load power factor. The most practical means of predicting the waveforms for Fig. 9-3 with other than resistive loads is by simulation.

9.3. ADDITIONAL REMARKS

In this chapter, the principle of ac phase control is discussed for single phase and for the wye-connected three-phase circuit without neutral connections. It is quite involved to determine the waveforms for this three-phase circuit, even for the case of purely resistive load. When the source and load neutrals are connected, or for the delta-connected arrangement including three single-phase circuits, as in Fig. 9-2, each connected across one pair of the input ac lines, the waveforms can be determined by single-phase analysis.

REFERENCES

1. Dewan, S. B. and Straughen, A. *Power Semiconductor Circuits*. New York: Wiley, 1975, Chapter Four.
2. Wood, Peter. *Switching Power Converters*. New York: Van Nostrand Reinhold, 1981, pp. 200–230.
3. Csáki, F. et al. *Power Electronics Problems Manual*. Budapest: Akadémiai Kiadó, 1979, pp. 176–209.

PROBLEMS

1. Consider the circuit in Fig. 9-2(a) with $R = 10 \ \Omega$, $L = 0$, and $v = 220\sqrt{2} \sin \omega t$.
   a) Carefully sketch the load voltage waveform for $\alpha = 60°$.
   b) Derive the general expression for $V_{o,e}$ as a function of V_m and α.
   c) Calculate $V_{o,e}$ for $\alpha = 0°$, $30°$, $60°$, $90°$, $120°$, $150°$, and $180°$.
   d) Plot the curve of $V_{o,e}$ versus α.
2. Calculate the input power factor for the circuit in Problem 1, with $\alpha = 30°$, $90°$, and $150°$.
3. Calculate the third and fifth harmonic components of the current as a percentage of the fundamental for the circuit of Problem 1, with $\alpha = 90°$.
4. Carefully sketch possible steady-state i_R waveforms for the circuit of Fig. 9-2(a) with
   a) $\alpha = 30°$ and $\omega L/R \gg 1$.
   b) $\alpha = 30°$ and $\omega L/R \ll 1$.
   c) $\alpha = 90°$ and $\omega L/R \gg 1$.
   d) $\alpha = 90°$ and $\omega L/R \ll 1$.
5. Plot v_{S-T} for the circuit in Fig. 9-3 with $\alpha = 15°$, $45°$, $75°$, $105°$, and $135°$.
6. Plot v_{T-R} for the circuit in Fig. 9-3 with $\alpha = 15°$, $45°$, $75°$, $105°$, and $135°$.

10
THYRISTOR CHOPPERS

10.1. ELEMENTARY CHOPPERS

With the advent of semiconductor devices which can operate in the kilohertz range as nearly ideal power switches, switching mode dc–dc voltage controls have become feasible. These approaches are very advantageous in a wide variety of applications because they provide a nearly lossless method of producing continuously adjustable output from a fixed dc voltage source. Various names have been used for these techniques, including time-ratio control, pulse modulation, switching mode regulators, and choppers. The latter name is used here because it is the designation most widely accepted when thyristor devices are employed. Important applications of thyristor choppers have included dc motor control for rapid-transit car and electric vehicle propulsion, excitation control for machines where faster transient response is required than achievable with phase-controlled rectifiers, and dc input voltage control for inverters.

Figure 10-1 shows an elementary form of the dc chopper. This is the switching circuit of Fig. 1-1, except switch S_1 is shown in place of the transistor switch. The elementary scheme of Fig. 10-1 is suitable only for supplying resistive loads where smooth output current is not required or for low-power applications where efficient filtering is not essential. Appreciable power losses result in filters containing linear circuit elements when such filters are used to smooth the output from the circuit in Fig. 10-1. In addition, the filter elements may become a significant part of the size, weight, and cost of the chopper when appreciable power is involved.

For the thyristor chopper, switch S_1 of Fig. 10-1 is a thyristor device which is cyclically opened and closed 1000 times per second or more. The fraction of the time that the switch is closed is varied to control the load voltage. The average load voltage V_L is related to the source voltage E_d as follows:

$$V_L = \frac{1}{T} \int_0^T V_L(t)\, dt = \frac{1}{T} \int_0^{t_{\text{CLOSED}}} E_d\, dt = \frac{E_d}{T} t_{\text{CLOSED}}$$

$$V_L = E_d \frac{t_{\text{CLOSED}}}{T_{\text{CLOSED}} + t_{\text{OPEN}}} \tag{10-1}$$

There are three basic operating modes for Fig. 10-1.

1. Pulse frequency modulation: t_{CLOSED} is held constant while the switching period T is varied.

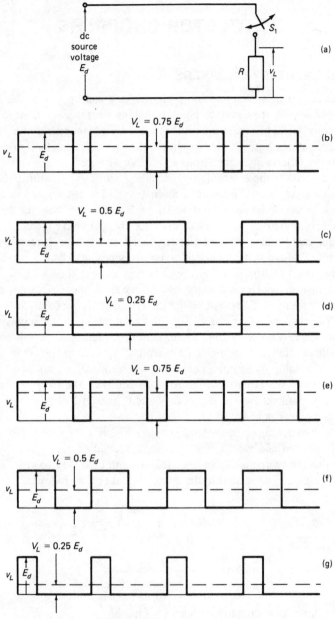

Fig. 10-1. Elementary dc chopper.

2. Pulse width modulation: t_{CLOSED} is varied while T is held constant.
3. A combination of pulse width and frequency modulation.

The waveforms of Fig. 10-1(b)–(d) illustrate pulse frequency modulation and pulse width modulation is shown in Fig. 10-1(e)–(g). In practice, some combination of these control modes often will result. It is quite difficult, particularly in simple choppers, to maintain the switch on-time constant over a wide frequency range including the effects of load changes and input voltage variations. However, pulse width controls with fixed frequency are quite easily achievable using modern digital IC logic techniques. In general, fixed-frequency variable pulse width circuits are somewhat more complex, but they have several important advantages; e.g., filtering is more efficient, the transient response is improved, and compensation to prevent closed-loop stability problems is somewhat easier.

Figure 10-2 shows an extremely important addition to the elementary chopper, which makes it possible to supply smooth dc to practical loads efficiently. Diode DF provides a path for the load current when $S1$ is open. This permits the use of filter inductance LF to provide sufficiently smooth dc load current

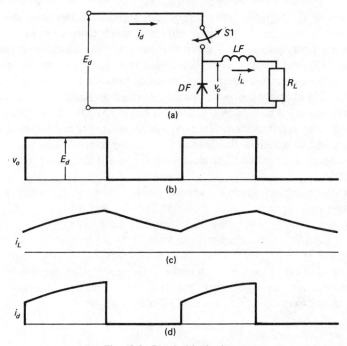

Fig. 10-2. Practical basic chopper.

for many applications. When the switching frequency is in the kilohertz range, a relatively small inductance is often adequate to reduce the ripple to a tolerable amount. For applications where only extremely low-amplitude ripple is permissible, more complicated L-C filter networks are generally added to the basic arrangement in Fig. 10-1. When the output is large enough to require devices in parallel, it is also possible to use chopper circuits in parallel, time displaced from one another, to achieve lower ripple. This is the "multicylinder" or "multiphase" approach referred to in Chapter 5.

Diode DF in the circuit of Fig. 10-2 permits filter inductance LF to provide the voltage transformation for the dc-to-dc conversion. Neglecting losses in switch $S1$ and assuming that the load current is continuous, the average output voltage across diode DF is related to E_d as given by Eq. (10-1).

Although it is difficult and possibly not of great importance to classify chopper schemes, some remarks along these lines are instructive. Three basic kinds of power circuits are generally used. These are shown in Fig. 10-3. A dc motor is used as the load since a source of power from the load is required for regenerative operation. Figure 10-3(a) is the simple nonreversing and nonregenerative circuit similar to Fig. 10-2. In this case, either a series or a separately excited machine is applicable. Often regenerative electrical braking is required, and this is possible with the approach of Fig. 10-3(b). In situations where both reversing and regeneration are required, a circuit of the form shown in Fig. 10-3(c) is necessary. This latter circuit is much more complicated than the basic chopper, particularly when the commutating circuit components are added. Thus, it has not often been used in vehicle propulsion applications. In these situations, reversing contactors provide fast enough reversal and appreciably simplify the power circuit. It must be noted that a separately excited machine must be used in Fig. 10-3(b) and Fig. 10-3(c) to provide regenerative and reversing operation. If a series machine is used, additional switching is required to maintain the field current in the same direction when the armature current is reversed to produce negative motor torque for reversing or regeneration.

Figure 10-4 illustrates another approach which can be used for reversing and regenerative braking in only one direction. This requires a double field, series dc motor, but it does simplify the SCR circuit to some extent.

In the remainder of this chapter, a number of SCR choppers are described. These are not necessarily the most advantageous circuits, but they illustrate a rather broad variety of possible approaches. The principal difference between the circuits discussed is the SCR commutation means. For this discussion, commutating circuits are classified into the following categories:

External impulse commutation
Resonant commutation (self-initiated)

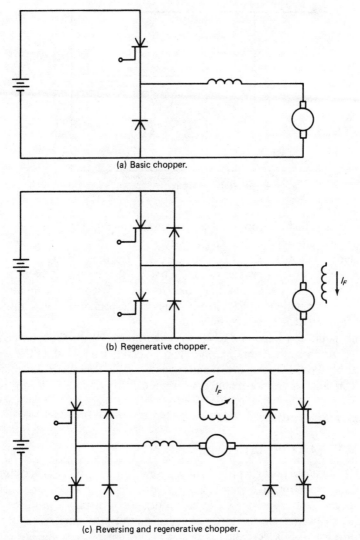

(a) Basic chopper.

(b) Regenerative chopper.

(c) Reversing and regenerative chopper.

Fig. 10-3. Choppers (commutating circuits not shown). (a) Basic chopper; (b) regenerative chopper; (c) reversing and regenerative chopper.

Resonant commutation (separately initiated)
Capacitor commutation

The first category involves the use of a power pulse transformer to couple a commutating pulse into the SCR circuit. This scheme has not been widely used

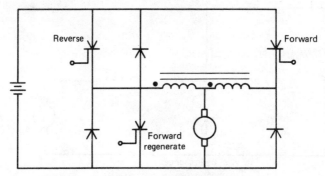

Fig. 10-4. Simplified reversing and regenerative chopper.

due to the facts that the power pulse transformer is a rather difficult component to design, usually relatively large, and also fairly expensive. In the resonant commutation schemes, an *L-C* circuit causes the current through the SCR to attempt to reverse. This process can be initiated either by gating the SCR to be commutated or by gating auxiliary devices as in McMurray commutation. Capacitor commutation implies situations where a capacitor is used to divert the current from an SCR to be commutated.

At this point it is important to consider the impact of the GTO on thyristor choppers. As these thyristors become more widely available, the SCR will be replaced by the GTO in many chopper applications. The usual commutating circuit components are not required with the GTO. However, it is still necessary to provide a freewheeling or coasting path for the load current when the chopper switch is opened.

10.2. MORGAN CIRCUIT [1]–[3]

One of the first practical chopper circuits is shown in Fig. 10-5. This chopper is referred to as the Morgan circuit in honor of its inventor. It is described here for historical and sentimental reasons, since Morgan is generally considered to be the father of power-semiconductor chopper controls. Another reason that the circuit in Fig. 10-5 is described is that it makes use of the saturable reactor. This component is a very rugged and reliable power switch. However, it is also generally expensive and often produces considerable acoustic noise.

In the circuit of Fig. 10-5, saturable reactor SR1 and capacitor *C*1 provide commutation of SCR1. For this form of commutation, the saturable reactor may be considered to operate as a switch. When it switches closed, it connects the commutation capacitor *C*1 across SCR1, thus providing an alternate path for the load current and a means of momentarily reversing the SCR anode-

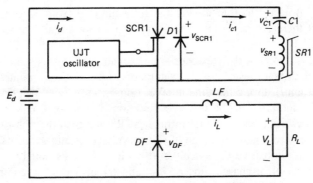

Fig. 10-5. Morgan circuit.

cathode voltage. With this mode of commutation, for a given supply voltage E_d, the SCR conduction angle or on-time interval is approximately constant as the unijunction transistor (UJT) frequency is varied to control the load voltage v_L. The period of time that the SCR is on is determined principally by the time required for the SR1 flux to move from positive saturation to negative saturation and return to positive saturation. The waveforms of Fig. 10-6 are believed to illustrate most clearly the saturable reactor–commutation principle. These waveforms assume ideal circuit components for the circuit of Fig. 10-5, negligible losses, negligible ripple in the load current, and an SCR on-time interval which is much greater than the commutation time interval required. At time t_0, it is assumed that capacitor $C1$ has an initial positive voltage across it equal to E_d, and the core flux of SR1 is assumed to be at positive saturation, $+\Phi_s$. When SCR1 is gated on at t_0, the capacitor voltage appears across SR1, driving its core flux toward negative saturation. Assuming that negligible exciting current is required by SR1, the capacitor voltage will remain essentially constant from time t_0 to time t_{1-2}. During this time interval, SCR1 is conducting load current I_L from the dc supply to the load. Diode DF is not conducting as an inverse voltage of E_d appears across this diode.

Time t_{1-2} is reached when the core flux of SR1 reaches negative saturation, $-\Phi_s$. At this time, capacitor $C1$ is discharged through SCR1 and the after-saturation inductance of SR1. This is actually a resonant discharge requiring a time $\pi\sqrt{L_{SR1}C1}$ seconds, assuming negligible losses, where L_{SR1} is the after-saturation inductance of SR1. However, the resonant discharge time of capacitor $C1$ is assumed very short relative to the time interval from t_0 to t_{1-2} for the waveforms in Fig. 10-6. Thus, the capacitor voltage drops very quickly, and is indicated by a vertical line at time t_{1-2} in Fig. 10-6(a). At the conclusion of the resonant discharge, the voltage across $C1$ will again be equal to E_d, but with the opposite polarity to that during the time interval from t_0 to t_{1-2}. It is assumed that there are negligible losses in the resonant circuit containing $C1$, L_{SR1}, and SCR1.

From time t_{1-2} to t_{3-6}, the capacitor voltage again remains constant while the flux in SR1 is driven toward positive saturation, $+\Phi_s$. SCR1 continues to conduct load current I_L during the time interval from t_{1-2} to t_{3-6}. When time t_{3-6} is reached, SR1 saturates in the positive direction. Capacitor $C1$ is again discharged in a resonant fashion, first through SCR1 and then through diode $D1$. After this second resonant discharge of $C1$, its voltage is returned to its original condition equal to positive E_d, since the losses have been assumed to be negligible. The resonant discharge time of L_{SR1} and $C1$ provides the commutating time required for SCR1. SCR1 is thus commutated off at time t_{3-6}. The capacitor voltage remains constant, and the load current circulates through diode DF until SCR1 is again gated on to start the next conduction interval.

The waveforms in Fig. 10-6 are obtained in practical circuits when the conduction interval of SCR1 is much longer than its commutating time requirement. These conditions usually exist when the maximum chopper frequency is 1000 Hz or less. The commutating-circuit losses in practical circuits are generally small enough so that capacitor $C1$ does not lose appreciable charge during its resonant discharging with the after-saturation inductance of SR1.

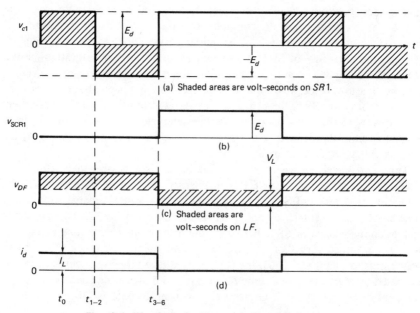

Fig. 10-6. Waveforms for Morgan circuit at low frequency.

The waveforms in Fig. 10-7 illustrate the behavior of the circuit of Fig. 10-5 when operating at higher chopping frequencies. In this case, the commutating time of the SCR may be an appreciable fraction of its conduction interval. These waveforms show the detailed operation during resonant discharging of capacitor $C1$. Saturable reactor SR1 is assumed to have a magnetization characteristic, as shown in Fig. 10-7(a). The load current is again assumed to have negligible ripple, and $D1$, and SCR1 are again assumed to have negligible forward drop when conducting, and negligible leakage when off. In Fig. 10-7, the losses in the commutating circuit are considered small but not, as they were in Fig. 10-6, negligible.

At time t_0, SCR1 is gated on to produce the waveforms in Fig. 10-7(b)–(i). The core flux in SR1 is assumed to be as shown in Fig. 10-7(a) at t_0. The initial charge on capacitor $C1$ is slightly greater than the dc supply voltage E_d. These initial conditions are chosen to produce steady-state operation over the cycle shown in Fig. 10-7.

After SCR1 is turned on at time t_0, $v_{SR1} = v_{C1}$. The flux of SR1 is driven toward negative saturation. There is no significant change in the capacitor voltage since the magnetizing current of SR1 is negligible. From t_0 to t_1, SCR1 conducts load current I_L, the full-supply voltage appears at the output across diode DF, and neither diode $D1$ nor DF is conducting. When the core flux of SR1 reaches negative saturation at time t_1, capacitor $C1$ is discharged through SCR1 and the after-saturation inductance of SR1. Because of the losses in the resonant L-C circuit formed by $C1$ and the after-saturation inductance of SR1, the capacitor voltage is somewhat less in magnitude at t_2 than it was at t_1.

As $C1$ discharges between t_1 and t_2, the energy $(1/2) C_1 V_{C1,m}^2$ is transferred to L_{SR1} in the form of $(1/2) L_{SR1} I_{C1,m}^2$ where L_{SR1} is the saturated inductance of SR1 and $I_{C1,m}$ is the peak value of the capacitor resonant-discharge current, as shown in Fig. 10-7(e). All the energy, except for losses, initially stored in $C1$ at time t_1 is transferred to L_{SR1} at the time between t_1 and t_2 when $v_{C1} = 0$ and $i_{C1} = I_{C1,m}$. As the current i_{C1} decreases again, the energy is returned to $C1$, reduced by the losses. Thus, at time t_2 the voltage across $C1$ is slightly lower in magnitude and opposite in polarity to its value at time t_1. Starting at time t_2, the flux in SR1 is driven toward positive saturation, reaching $+\Phi_s$ at time t_3. Again, during the interval from t_2 to t_3, the capacitor voltage is constant, assuming negligible exciting current for SR1. The volt-second area A_{2-3} is that required to drive SR1 from $-\Phi_s$ to $+\Phi_s$.

At time t_3, a second resonant oscillation of capacitor $C1$ with the after-saturation reactance of SR1 is initiated. When the capacitor current reaches a value equal to the load current I_L, SCR1 stops conducting, and diode $D1$ begins to conduct. The resonant oscillation of $C1$ continues, otherwise, the same as during interval t_1 to t_2, until time t_4. At this point, the capacitor current has reduced back to a value equal to the load current I_L; diode $D1$ blocks; forward

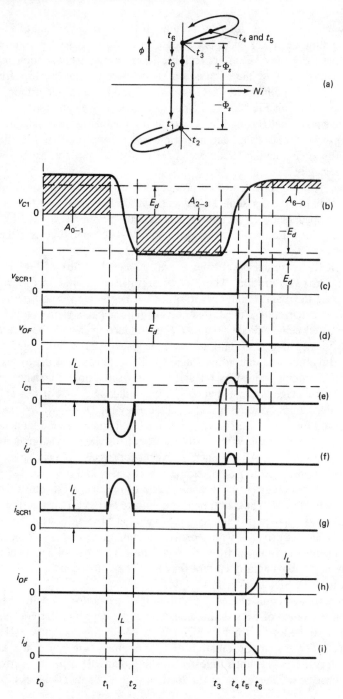

Fig. 10-7. Waveforms for Morgan circuit at high frequency.

voltage is reapplied to SCR1; and the capacitor continues charging from the dc supply with a constant current equal to the load current I_L during the interval from t_4 to t_5. When time t_5 is reached, the capacitor has been charged to E_d, the reverse voltage across diode DF reaches zero, and diode DF begins conduction. The capacitor receives its final charge during the interval from t_5 to t_6. This is a continuation of the resonant oscillation of $C1$ and the after-saturation inductance of SR1, but now the capacitor current i_{C1} comes from the dc supply, as both SCR1 and $D1$ are blocking. When the oscillation has proceeded to the point at which the current i_{C1} reaches zero, diode DF is conducting the full load current I_L. The voltage across SR1 is again in the direction to drive its core flux toward negative saturation. The capacitor voltage remains approximately constant after t_6, resetting SR1 to the point t_0 in Fig. 10-7(a) to begin the next cycle of operation.

The time interval t_0 to t_1 is a function of the interval from t_6 to t_0, as SR1 is partially reset from $+\Phi_s$ between time t_6 and the beginning of the next conducting interval. The remainder of the resetting of SR1 to $-\Phi_s$ occurs between t_0 and t_1. For this reason, the SCR-conducting period is reduced somewhat as the UJT frequency is reduced.

The time interval from t_3 to t_4 is a function of the load current I_L, as shown in Fig. 10-7(e). Thus, as the load current is increased, the interval of time during which the supply voltage E_d is impressed on the load circuit is slightly reduced.

It is interesting to note that the circuit in Fig. 10-5 is practically self-regulating to changes in supply voltage. When the supply voltage E_d is increased, SR1 is driven over its flux range in less time. Thus, the SCR-conduction interval is reduced when the dc supply voltage E_d is increased. For a fixed UJT frequency, this means that there is a compensating effect for supply-voltage fluctuations.

10.3. SERIES RESONANT CHOPPER [4]

In the circuit of Fig. 10-8, the commutating inductor and capacitor are in series with the SCR. Again the SCR acts as a switch which alternately closes and opens to provide time ratio control of the output voltage to LF and the load.

Figure 10-9 illustrates the steady-state waveforms, assuming negligible commutating losses, ideal components, and negligible ripple in the load current. Assuming steady-state conditions, just prior to closing the SCR1 switch, the load current is coasting or freewheeling through diode DF, inductor LC, inductor LF, and the load R_L. The capacitor voltage is constant and equal to zero. When SCR1 is gated, its anode-cathode voltage immediately goes to zero and v_o abruptly jumps to E_d. For a short time interval, the voltage E_d is impressed on inductor LC, which causes the current through LC to fall linearly toward zero. As this current decreases, the SCR forward current increases. When the SCR current becomes equal to the load current, resonant charging of the capacitor is initiated. The current through LC and C is sinusoidal, considering negligible loss in the charging circuit. At the end of the first half-cycle of the

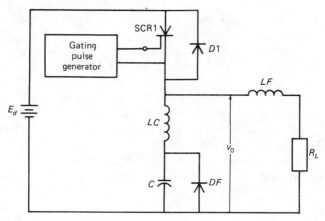

Fig. 10-8. Series resonant chopper.

resonant charging interval, the capacitor voltage is approximately twice the dc source voltage, and the current returns to zero. On the second half-cycle the LC-C current is reversed. When it reaches the same magnitude as the load current, the SCR current is zero. As the second half-cycle of resonance continues, diode $D1$ conducts and current flows back into the positive dc source line. When the LC-C current oscillates back down to the load-current level, the SCR-$D1$ branch opens. From this point onward, the commutating capacitor discharges into the load circuit. Assuming that inductor LF is large enough to maintain essentially constant load current, the capacitor voltage decreases at almost a constant rate until it reaches zero. When the voltage across C is zero, diode DF conducts. The path for current is now through DF, LC, LF, and the load. When the SCR is gated to start the next on-interval, there is again a brief period during which diode DF continues to conduct. The dc source voltage appears across LC, causing the current through LC to fall linearly to zero. From this time onward, the same sequence of events occurs as took place during the first on-interval.

The unique feature of the circuit in Fig. 10-8 is that the capacitor voltage is reset to zero at the beginning of each on-interval. This assumes that continuous current flows in the load, which will be the case over a wide load range. When the load is such that the capacitor voltage has not reached zero at the start of the next on-interval, the commutating capability is reduced. For normal operation, the peak value of the LC-C current should be 1.5 to 2.0 times the load current. This provides an SCR turn-off time approaching the half-period of the LC-C resonant frequency. The period of conduction of diode $D1$ is the turn-off time provided for the SCR.

For a wide range of loads, the capacitor voltage is reset to zero prior to the beginning of each on-interval as a result of the action of the series resonant cir-

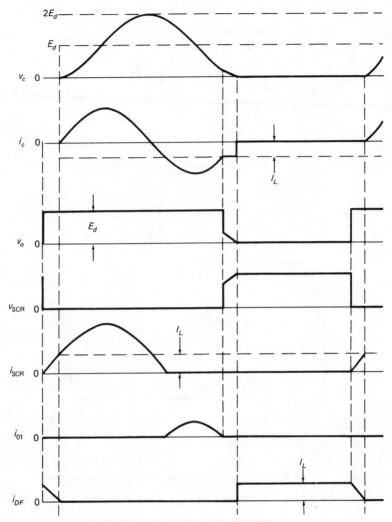

Fig. 10-9. Waveforms for Fig. 10-8.

cuit and the diode in parallel with the capacitor. This produces a fixed commutating voltage for a variety of loads.

10.4. SINGLE AUXILIARY SCR CHOPPER

This is another very early chopper circuit first reported in the literature in 1963 by Heumann [5]. It has been quite widely used since that time.

Figure 10-10 shows the basic circuit, and the waveforms are illustrated in Fig. 10-11. For simplicity, it is assumed that there is negligible ripple in the

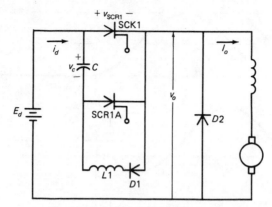

Fig. 10-10. Single auxiliary SCR chopper.

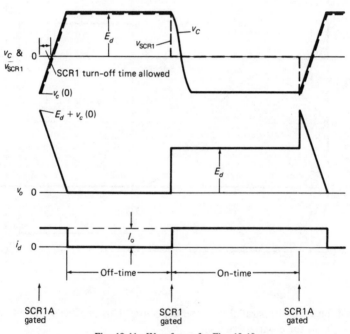

Fig. 10-11. Waveforms for Fig. 10-10.

motor current and that the chopper is operating in steady state. When SCR1A is fired, capacitor C charges to E_d volts. Its initial voltage is assumed to be negative with a magnitude somewhat smaller than E_d. When the capacitor voltage reaches E_d, the motor current abruptly begins "coasting" through diode $D2$.

This action continues throughout the chopper off-time. When SCR1 is gated, the motor current is immediately supplied by the dc source. In addition, the commutating capacitor C is discharged in a damped sinusoidal fashion. Assuming finite, but small, losses in the SCR1-D1-L1-C circuit, the capacitor voltage reverses to a slightly lower magnitude negative voltage than E_d. Since diode D1 prohibits reverse current flow in capacitor C when SCR1A is off, the circuit will remain in this state for the duration of the on-time. Then when SCR1A is gated again, the process is repeated.

10.5. BRIDGE AUXILIARY SCR CHOPPER [6], [7]

Figure 10-12 shows the circuit and Fig. 10-13 its waveforms. For simplicity these waveforms assume steady-state operation and negligible ripple in the motor current. SCR1A and SCR2A are gated simultaneously, charging capacitor C to E_d volts. When the capacitor voltage attempts to increase beyond E_d, diode D begins to conduct, abruptly causing the current from the dc source to drop to zero and the load current to "coast" thru diode D. When SCR1 is gated, the load current is supplied from the dc source and diode D is reverse-biased. SCR3A and SCR4A are gated next to comutate SCR1. The commutation process is the same as when SCR1A and SCR2A were gated. The unique advantageous feature of this chopper scheme is that the capacitor is left in the state ready for the next commutation after each commutation of the main SCR. This results in a high-efficiency circuit. However, the price paid for this feature is the multiplicity of auxiliary SCRs.

10.6. MODIFIED McMURRAY COMMUTATED SCR CHOPPER

This is an example of a regenerative chopper, as shown in Fig. 10-3(b). The configuration considered is essentially one-half of the original McMurray single-

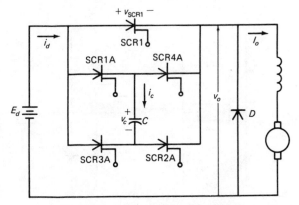

Fig. 10-12. Bridge auxiliary SCR chopper.

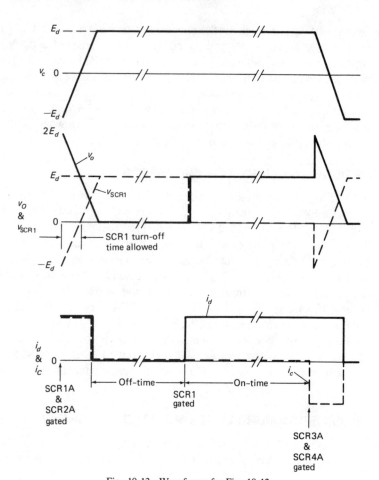

Fig. 10-13. Waveforms for Fig. 10-12.

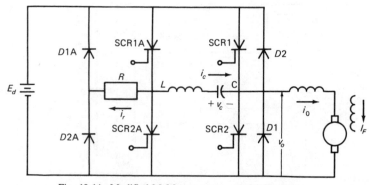

Fig. 10-14. Modified McMurray commutated SCR chopper.

phase bridge inverter, except with the addition of two diodes and a resistor. Figure 10-14 is the circuit [8].

It is important to note that the reversal of power flow in the circuit of Fig. 10-14 is accomplished by reversing the direction of current flow. This is different from the situation in the phase-controlled rectifier or line-commutated inverter. In that case, the reversal of power flow is achieved by reversing the polarity of the dc circuit voltage, and the dc circuit current always flows in the same direction. This means that if the load in Fig. 10-14 is a series dc motor, either the field or the armature must be reversed during regeneration in order to produce a braking torque. However, if a separately excited dc machine is used and the inductance L_M is a separate inductor, then braking torque automatically is produced during regeneration since the armature current is reversed.

For simplicity it will be assumed that the motor current remains essentially constant during commutation. This is true in most practical cases, since there is generally sufficient motor inductance to prevent significant change in the motor current during the relatively short commutating intervals.

10.6.1. Motoring

Figure 10-15 illustrates the waveforms for motoring operation. It is assumed that SCR1 and SCR2A are both gated at t_0 with zero initial capacitor voltage, with no initial current in the commutating inductance or the load inductance, and with the motor at rest so its CEMF is zero. In addition, the CEMF will be considered negligible for the cycle of operation shown. This would be true in a practical case, since for any reasonably large motor, its mechanical time constant would be such that the speed would not change significantly during a few chopping periods, assuming a chopper frequency of 1000 Hz or more.

The waveforms of Fig. 10-15 can be understood most clearly by referring to the "active" circuits shown in Fig. 10-16.

t_0–t_1. SCR1 and SCR2A are simultaneously gated to start this interval. Two currents flow through SCR1 — the motor current and the commutating L-C current. The commutating current is a negative half-sinusoid, and the capacitor voltage swings negatively to twice the dc supply voltage, considering negligible loss in the E_d-SCR1-C-L-SCR1A circuit.

t_1–t_2. This is the major portion of the SCR1 on-time. Since the capacitor voltage is greater in magnitude than the dc source voltage, diode $D2A$ conducts a relatively small current, causing the capacitor to discharge down to E_d. The resistance R is assumed to be large enough so that this is an overdamped discharge such that the capacitor voltage decays exponentially to negative E_d volts. Also, during this interval, the motor current i_0 builds up to the value indicated.

t_2–t_3. SCR1A is gated at time t_2 to commutate SCR1. The same type of resonant discharge occurs as during the t_0–t_1 interval. However, when the capacitor

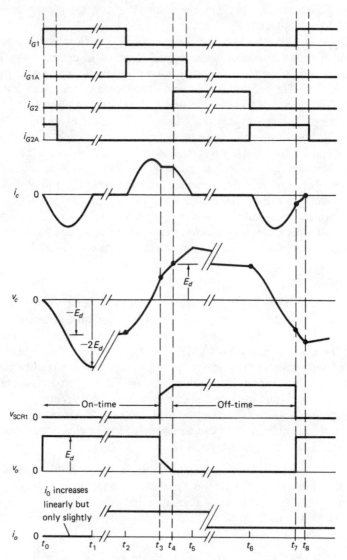

Fig. 10-15. Waveforms for motoring operation of Fig. 10-14.

current oscillates back down to a value equal to the motor current, the SCR1-D2 branch opens.

t_3–t_4. After the SCR1-D2 branch opens, there is a period during which the capacitor continues charging, essentially linearly, assuming the nearly constant

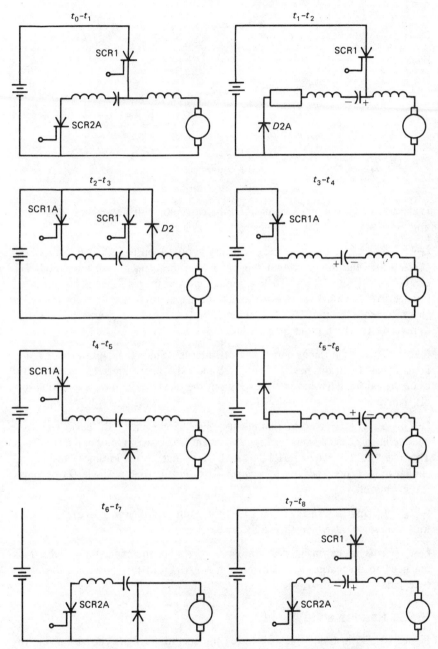

Fig. 10-16. Sequential active circuits for waveforms of Fig. 10-15.

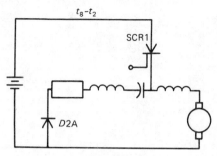

Fig. 10-16. (continued)

load current. This interval ends when the capacitor voltage reaches E_d. At this time, coasting diode $D1$ becomes forward-biased.

t_4-t_5. The capacitor completes charging via the loop including E_d-SCR1A-L-C-$D1$. Since the motor current is greater than the capacitor current, diode $D1$ can be considered a short. That is, the net current in this diode is always positive during this interval, even though the capacitor current is considered a "back-current" through $D1$. This interval ends when the capacitor current tries to reverse, at which point SCR1A blocks.

t_5-t_6. This is the major part of the off-interval. Since the capacitor voltage is greater than E_d, it discharges through R and $D1A$ approximately exponentially as described for interval t_1-t_2. In addition, the motor current decays during this off-interval.

t_6-t_7. SCR2A is gated to commutate SCR2. However, SCR2 never was conducting in this case because the motor current was coasting through diode $D1A$. This interval ends when SCR1 is gated. Note that in this instance, an interval similar to t_3-t_4 does not occur because SCR1 is fired before diode $D1$ becomes reverse-biased.

t_7-t_8. During this interval the capacitor continues charging until its current tries to reverse, at which time SCR2A blocks.

t_8-t_2. This is the on-interval similar to t_1-t_2, but the capacitor voltage does not need to discharge as much through $D2A$. From this point on, the cycle is repeated.

10.6.2. Regenerating

Figure 10-17 illustrates the waveforms for operation of the chopper in the regenerative mode. For this case it is assumed that the motor is separately excited and that inductor L_M is large enough so that i_0 again can be considered constant

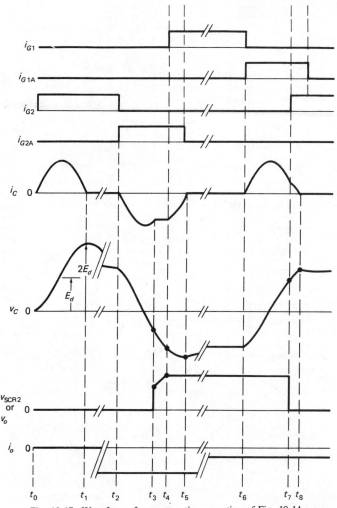

Fig. 10-17. Waveforms for regenerative operation of Fig. 10-14.

during the commutation intervals. SCR2 and SCR1A are simultaneously gated at t_0 with zero initial capacitor voltage and no initial current in the commutating inductor or load inductor, but with a finite value of CEMF on the motor. In other words, the motor is assumed to be freewheeling at a fixed speed at t_0. Figure 10-18 shows the "active" circuits during each time interval.

10.7. ADDITIONAL REMARKS

The circuits which have been discussed by no means exhaust the possibilities. However, they do illustrate a variety of approaches.

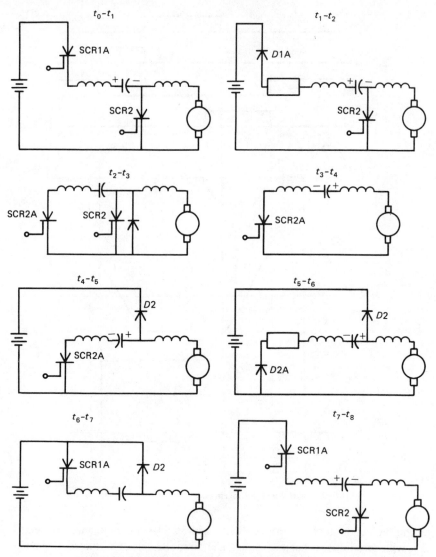

Fig. 10-18. Sequential active circuits for waveforms of Fig. 10-17.

Another chopper approach which has been used quite widely in Japan and Europe involves the use of reverse-conducting thyristors. These simplify the chopper because of the "built-in" reverse-parallel diodes.

Multiphase choppers can be used to reduce the input and output filtering by increasing the equivalent chopper frequency.

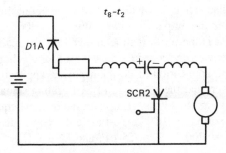

Fig. 10-18. (continued)

REFERENCES

1. Morgan, R. E. A New Magnetic-Controlled Rectifier Power Amplifier with a Saturable Reactor Controlling On Time. AIEE Transactions 80, Part I: 1961, pp. 152–155.
2. Morgan, R. E. Magnetic Silicon Controlled Rectifier Power Amplifier. U.S. Patent 3, 019, 355, January 30, 1962.
3. Bedford, B. D. and Hoft, R. G. *Principles of Inverter Circuits*. New York: John Wiley & Sons, 1964, chap. 10.
4. Hoft, R. G., et al. Thyristor Series Resonant DC-DC Chopper. IEEE Transactions on Magnetics MAG-8, no. 3: September 1972, pp. 286–288.
5. Heumann, K. Pulse Control of DC and AC Motors by Silicon Controlled Rectifiers. Washington, DC: IEEE Proceedings of Intermag Conference, April 17–19, 1963, pp. 5-1-1 through 5-1-13.
6. Wouk, Victor. High Efficiency, High Power, Load Insensitive DC Chopper for Electronic Automobile Speed Control. Power Semiconductor Applications, Vol. I: General Considerations, Edited by John D. Harnden, Jr., and Forest B. Golden, IEEE Press 1972, pp. 393–402.
7. Heumann, K. and Stumpe, A. C. *Thyristoren Eigenschaften und Anwendungen*. Stuttgart: B. G. Teubner, 1970.
8. Dewan, S. B. and Straughen, A. *Power Semiconductor Circuits*. New York: John Wiley & Sons, 1975, pp. 346–352.

PROBLEMS

1. Consider the basic chopper of Fig. 10-2(a) with $E_d = 250$ V, $R = 1 \ \Omega$, $LF = 50 \ \mu$H, an on-time of 50 μs, and a switching period of 100 μs. Without solving differential equations, sketch an accurate approximation to the steady-state load current (i_L) waveform.

2. Assume that inductor LF is large enough so that there is negligible ripple in the load current for the series resonant chopper shown in Fig. 10-8. With $E_d = 500$ V and a steady-state load current of 100 A, calculate the values of LC and C required to provide 50 μs turn-off time for SCR1 and a peak capacitor current of twice the steady-state load current.

3. The bridge auxiliary chopper of Fig. 10-12 is operating in steady state with negligible ripple in the output current I_0. With $I_0 = 100$ A, $E_d = 250$ V and $C = 50$ μF, what is the turn-off time allowed for the main SCR (SCR1)?

4. Describe a proper start-up procedure for the McMurray chopper of Fig. 10-14 with regard to the application of dc input power, application of control and logic circuit power, and the sequence of SCR gating signals.

5. For the circuit of Fig. 10-14 operating with the on-time equal to the off-time and a switching period of 1000 μs, estimate the power dissipated in resistor R, given that $E_d = 200$ V, $C = 25$ μF, and $R = 5$ Ω.

11
SELF-COMMUTATED THYRISTOR INVERTERS

11.1. INVERTER CLASSIFICATION

The classification philosophy used in this text divides all inverters into two classes — self-commutated inverters and externally commutated inverters. A self-commutated inverter is one in which the means of commutation is included within the inverter itself. This is the case for circuits including electronic power switches with turn-off capability. Thus, a bipolar transistor, GTO thyristor, or power FET inverter is self-commutated. In addition, SCR inverters which include capacitors, inductors and possibly auxiliary thyristors to provide commutation are also examples of self-commutated inverters.

Externally commutated inverters are those in which the commutation means is external to the inverter circuit. The phase-controlled rectifier operating as a line-commutated inverter is the best known externally commutated circuit. Circuits in which the nature of the load provides the means of commutation are also examples, i.e., an SCR inverter supplying a synchronous motor where commutation is accomplished by the motor CEMF, or an inverter in which a particular load filter or a resonant circuit is used to cause commutation. Another example of an externally commutated inverter is one in which the dc bus is interrupted to produce commutation. Although such an approach may be considered self-commutated if the electronic switch hardware for causing the bus interruption is in the same package as the inverter switches, this scheme is functionally an externally commutated approach.

Self-commutated circuits are further subdivided into voltage-source inverters and current-source inverters. As the names imply, the VSI is fed from a voltage source, and the CSI is fed from a current source. The input for the simplest VSI is a battery. However, the dc source voltage also may be adjusted to control the inverter output voltage. In any case, the VSI is such that the dc input voltage remains relatively constant in spite of large transient or steady-state variations in the input current. For the CSI, the input current remains relatively constant over a wide range of dc input voltage variations. The most common input for the CSI is a phase-controlled rectifier supplying the inverter in series with a relatively large inductor. This forces a given input current to flow at least throughout a cycle of operation, in spite of the switching which occurs within the inverter circuit.

In this chapter, one VSI and one CSI are discussed in some detail to illustrate the behavior of these inverters.

11.2. McMURRAY INVERTER

When the SCR first became available, there was considerable effort to develop inverter circuit techniques which would utilize the unique features of this electronic power switch. The most successful and widely used SCR inverter is the circuit devised by McMurray [1], [2]. One of the simplest circuit configurations is shown in Fig. 11-1. This arrangement is also the basic building block for the single-phase and three-phase bridge versions of the McMurray inverter.

11.2.1. Half-Bridge—No Load

The operation of the circuit in Fig. 11-1 is as follows. Assuming the correct initial capacitor voltage, a main SCR is commutated by gating the appropriate auxiliary. When SCR1A or SCR2A is triggered, this causes a resonant pulse of current to flow in opposition to that being carried by the conducting main SCR. When the main SCR current has been reduced to zero, the commutating-current pulse will continue to flow in the appropriate feedback or bypass diode, thus providing a small reverse voltage across the main SCR for the time interval necessary for it to regain its blocking capability.

One of the most advantageous features of this inverter is that after a given main SCR is commutated, the capacitor voltage is left at a value required to commutate the next main SCR. This feature combined with the pulse nature of the commutating action provides a highly efficient inverter for operation at normal power frequencies. In fact, an efficiency of over 90 percent can be obtained, with SCR turn-off times in the 20-μs range, for circuit operating frequencies up to several kilohertz.

Another very advantageous feature of this inverter approach is that with the proper SCR triggering sequence, sufficient commutating capacitor voltage is assured for a wide range of load conditions.

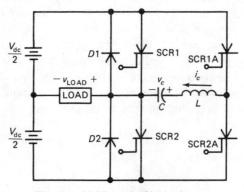

Fig. 11-1. McMurray half-bridge inverter.

The detailed operation of Fig. 11-1 is illustrated in Fig. 11-2. It is assumed that the load current is negligible but that the main SCRs do conduct when they are triggered. Ideal semiconductor devices are also assumed, and the losses in the commutating circuits are considered negligible. In actual practice, the load would need to be sufficient for the main SCR holding current to flow. However, with the assumption of ideal semiconductors, the holding current can be considered negligible. In modern devices it is generally much less than 1 percent of the rated forward current.

Figure 11-3 indicates the equivalent circuit during each interval of operation illustrated in Fig. 11-2. At time t_0, SCR1A is gated to commutate SCR1. From time t_0 to t_1, two current loops exist. The resonant current through the inductance and capacitance flows in diode $D1$, maintaining zero potential across the SCR1-$D1$ branch until the end of the resonant current positive half-cycle. During the entire positive half-cycle of $i_c(t)$, the current through SCR1 is zero

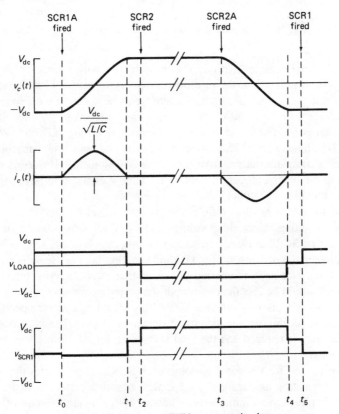

Fig. 11-2. No load — negligible commutating loss.

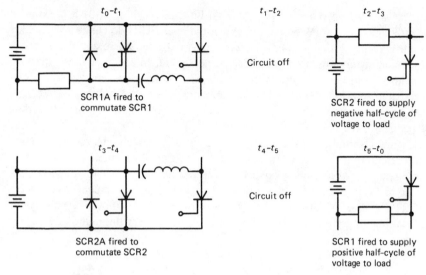

Fig. 11-3. Equivalent circuits during intervals shown in Fig. 11-2.

and in the practical case, diode $D1$ provides a small reverse voltage across SCR1. For this situation, the turn-off time provided for SCR1 is an entire half-cycle of the commutating circuit resonant frequency, i.e., $\pi\sqrt{LC}$ seconds. The load voltage remains equal to $V_{dc}/2$ during time interval t_0–t_1, since the SCR1-$D1$ branch is shorted because of the current flow through $D1$. At time t_1, the SCR1-$D1$ branch opens. The current through the commutating capacitor tries to reverse to complete the negative half-cycle of resonance. However, diode $D1$ blocks in the reverse direction, and SCR1 is now assumed to have regained its forward-blocking capability.

From time t_1 to t_2, the entire circuit is open. That is, there are no currents flowing, resulting in zero load voltage and $V_{dc}/2$ forward volts across SCR1.

At time t_2, SCR2 is gated, producing the negative half-cycle of load voltage. The slanted pairs of lines in Fig. 11-2 indicate that the time scale is compressed during t_2–t_3. Normally the period of the resonant frequency for the commutating circuit would be less than one-tenth the period of the inverter operating frequency. With 20-μs turn-off time SCRs, this would be true for operation up to on the order of 2500 Hz. If the circuit operating frequency is increased further, the efficiency is reduced and the load voltage no longer closely approximates a square wave.

At time t_3, SCR2A is gated to commutate SCR2. This initiates the same type of resonant action as during t_0–t_1, but with negative current, which resets the capacitor voltage to its initial value. Interval t_0–t_5 is again a circuit off-interval. At t_5, SCR1 is gated to start the next cycle of operation.

11.2.2. Half-Bridge—Resistive Load

Figure 11.4 is a sketch of the waveforms with the same assumptions as for Fig. 11.2 except that a finite resistive load is now assumed. The load is such that the current is equal to half the peak resonant current through the commutating LC branch, i.e.,

$$\frac{V_{dc}/2}{R} = \frac{1}{2}\frac{V_{dc}}{\sqrt{L/C}} \tag{11-1}$$

The operation is the same with this resistive load, except that an additional interval occurs where a different equivalent circuit exists. From t_0 to t_1, the equivalent circuit is as shown in Fig. 11-3 for interval t_0–t_1. However, at time

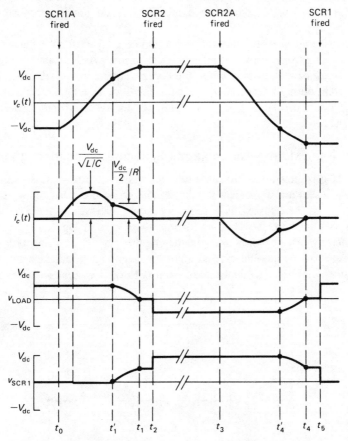

Fig. 11-4. R-load—negligible commutating loss.

t_1', the commutating LC current has returned to a value equal to the load current. Thus, at this instant the current through the SCR1-D1 branch is zero, and this branch opens. From t_1' to t_1 in Fig. 11-4, the equivalent circuit is as shown in Fig. 11-5. It is assumed that the resistive load is such that the current still oscillates to zero following a damped sinusoid during interval $t_1'-t_1$. At time t_1, the circuit is again open and the remaining operation is the same as for the no-load case.

It should be noted that steady-state operation is not shown in Fig. 11-4. The capacitor voltage is greater in magnitude after commutation of SCR1 than it was initially. With the assumption of negligible commutating circuit losses, this buildup of capacitor voltage continues on subsequent commutations. However, in a practical situation, commutating losses prevent a continual buildup. The capacitor voltage would be greater in magnitude than V_{dc} volts just prior to commutation. It would then lose voltage during commutation because of losses, which would be just recovered by the additional capacitor charging during $t_1'-t_1$.

It also should be noted that the turn-off time allowed for the main SCRs is reduced going from no load to a finite resistive load. As indicated in Fig. 11-4, SCR1 is reverse-biased only while the commutating current is greater than the load current during interval t_0-t_1'. For the condition assumed in Eq. (11-1), the turn-off time allowed is from 30° to 150° of the resonant half-cycle, or 2/3 $\pi\sqrt{LC}$ seconds.

11.2.3. Half-Bridge—No Load, but Including Commutating Losses

Figure 11-6 is intended to illustrate two important features—the effect of commutating losses and a means to make up charge on the capacitor which is removed due to these losses. For simplicity the no-load case is again considered. The assumptions are the same as for the waveforms in Fig. 11-2, except that commutating circuit losses are considered, the main SCR firing is advanced with respect to the auxiliaries, and a greater magnitude initial capacitor voltage is considered. The equivalent circuit from t_0 to t_1'' is as for t_0-t_1 in Fig. 11-3.

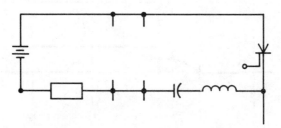

Fig. 11-5. Equivalent circuit during interval $t_1'-t_1$.

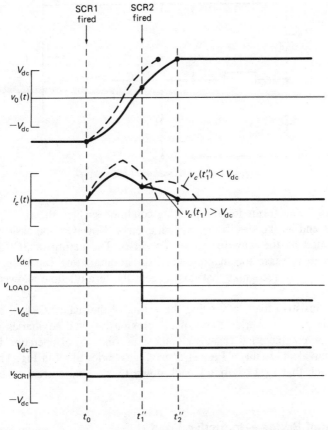

Fig. 11-6. No load—with commutating loss and main SCR firing advanced.

However, $v_c(t)$ and $i_c(t)$ do not follow exactly sinusoidal waves because of losses. Thus, the capacitor voltage would be lower in magnitude at the end of the first resonant current half-cycle. During each subsequent commutation, the capacitor voltage would be reduced in magnitude until the circuit failed to commutate. In order to prevent this, the main SCR firing is advanced. Thus, SCR2 is gated at t_1'' in Fig. 11-6. This produces a new interval of operation from t_1'' to t_2'' with the equivalent circuit shown in Fig. 11-7.

During the interval t_1''–t_2'', the capacitor voltage is changed by the combined effects of the net driving voltage $[V_{dc} - v_c(t_1'')]$ and the initial inductor current, $i_c(t_1'')$. For the steady-state condition illustrated in Fig. 11-6, the capacitor voltage would just reach the magnitude it had started with at t_0, by the time the current $i_c(t)$ tried to reverse at t_2''. The remaining intervals of operation are similar to those shown in Fig. 11-3.

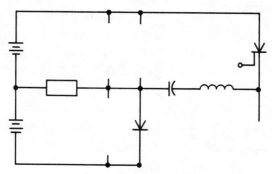

Fig. 11-7. Equivalence circuit during $t_1''-t_2''$.

The exact waveforms for an operating condition such as shown in Fig. 11-6 would depend on the specific circuit parameters. However, the basic principle is illustrated by the waveforms of Fig. 11-6. The gating of SCR2 can be advanced to increase the magnitude of the commutating capacitor voltage. It is important to recognize that this reduces the turn-off time allowed for the main SCR.

With a resistive load, advancing the gating of the main SCRs so that their conducting periods overlap those of the previously gated auxiliaries will also increase the commutating capacitor voltage. An interval of operation indicated by the equivalent circuit of Fig. 11-5 may occur prior to t_1'' in Fig. 11-6 if the load is such that $i_c(t)$ oscillates back down to the load-current level before SCR2 is gated.

11.2.4. Half-Bridge—Inductive Load

Figure 11-8 illustrates the waveforms for an inductive load where it is assumed that the inductance is such that there is negligible change in the load current during the commutating intervals. This is a practical assumption for a wide range of inductive loads at normal power frequencies, since the commutation intervals are short. Again, ideal semiconductors are assumed and the commutating circuit losses are neglected for simplicity. This results in an increasing magnitude of capacitor voltage on successive cycles of inverter operation. In practical situations, the circuit losses will limit the buildup of capacitor voltage.

Figure 11-9 shows the equivalent circuits during each time interval indicated on Fig. 11-8. For the initial capacitor voltage shown and the assumption of negligible commutating losses, the operation is somewhat different on the second half-cycle than on the first. With an initial capacitor voltage of in the order of V_{dc} or less, interval t_1-t_2 exists. At time t_1, $i_c(t)$ has oscillated back down to the load current.

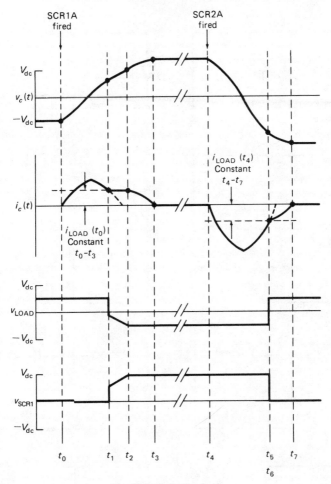

Fig. 11-8. Inductive load.

Then there is a period of operation with the equivalent circuit, as in Fig. 11-9 for t_1-t_2. Since the load current is assumed approximately constant during commutation, the commutating capacitor charges linearly during t_1-t_2. Also, the voltage across the commutating inductance is nearly zero during this interval because the assumption of negligible change in the load current during commutation implies that the commutating inductance is small with respect to the load inductance. At time t_2, the capacitor voltage reaches V_{dc} volts. This is also the voltage across SCR1, and thus the load voltage is now negative $V_{dc}/2$ volts. When $v_c(t)$ attempts to rise further, diode $D2$ conducts producing the equivalent circuit in Fig. 11-9 for t_2-t_3.

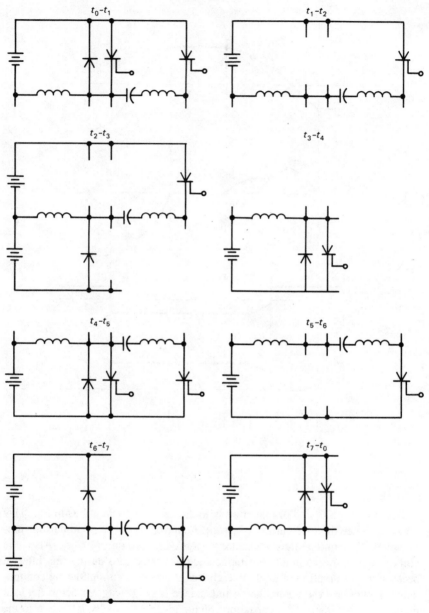

Fig. 11-9. Inductive load.

During interval t_2-t_3, two current loops exist. The nearly constant load current flows with a voltage of negative $V_{dc}/2$ volts across the load. In addition, the commutating L-C continues its resonant action with V_{dc} driving volts, V_{dc} initial capacitor volts, and an initial inductor current at t_2 equal to the load current. Interval t_2-t_3 ends when the commutating current attempts to reverse, since this will cause SCR1A to go off.

During t_3 to t_4, the load voltage is negative $V_{dc}/2$ volts. This causes the inductive load current to decrease and to subsequently go through zero at a point determined by the load power factor. When the load current reverses, diode $D2$ blocks and SCR2 now conducts, which results in the same equivalent circuit. It is important to note that SCR2 must be gated at the point where the load current reverses. For operation with a wide range of loads, the main SCRs are usually gated for essentially the entire half-cycle, during which they may be expected to conduct.

When SCR2A is gated to commutate SCR2, a similar sequence of events is initiated. However, as noted in Fig. 11-8, the capacitor voltage may now be greater in magnitude than V_{dc} when time t_5 is reached. This means the interval t_5-t_6 does not exist, and the equivalent circuit changes from that during t_4-t_5 to that for t_6-t_7 in Fig. 11-9. From t_7 to t_0, the load current is reversed and then the sequence of events is repeated.

It should be emphasized that the waveforms in Fig. 11-8 illustrate the operation for a wide range of inductive loads — from a purely inductive load to a nearly unity power factor load, but with a series load inductance still in the order of ten times the commutating inductance.

11.2.5. Half-Bridge — Capacitive Load

The McMurray inverter essentially delivers a constant-voltage square-wave output. Thus, when a capacitive load is supplied, it is necessary to include series impedance to limit the current during voltage switching. For this brief discussion, it will be assumed that there is an inductance in series with the capacitive load. The power factor of the total load will be assumed leading. However, the inductance will be sufficient to limit the load current during load voltage switching. In fact, in many practical cases, the series inductance in the load circuit may be sufficient to maintain the current constant during commutation intervals, as was assumed in the inductive load case. This merely requires an inductance in series with the load, or as a part of the load, which is in the order of ten times larger than the commutating inductance. In its simplest form, this would imply a series L-C load with sufficient inductance but still having a leading power factor at the inverter operating frequency.

The operation of the circuit of Fig. 11-1 with a capacitive load is very similar to the no-load operation. Assuming that SCR1 has been on for nearly a half-cycle, with a leading load, the load current would have already reversed. Thus, at the time when SCR1A is gated to commutate SCR1, diode $D1$ is conducting. SCR1 is already off and its commutation is unnecessary. However, when SCR1A is gated, the commutating capacitor voltage is reversed so that it is ready to commutate SCR2 if that becomes necessary. Assuming that $D1$ is conducting when SCR1A is gated, the entire half-cycle of resonance occurs with the equivalent circuit, as during $t_0–t_1$ in Fig. 11-3. Diode $D1$ maintains the SCR1-$D1$ branch shorted during this interval. When the commutating capacitor current attempts to reverse, SCR1A goes off. The load current then continues to flow in $D1$ until SCR2 is fired. Normally, SCR2 is fired somewhat before SCR1A goes off. This will transfer the leading current in the load immediately to SCR2, and it will provide further charging of the commutating capacitor to make up losses, as discussed in Section 11.2.3.

With the leading load, the load current reverses before the end of the next cycle. Thus, diode $D2$ is conducting when SCR2A is gated to commutate SCR2. Since SCR2 is already off, its commutation is unnecessary. However, gating SCR2A reverses the capacitor voltage so that it is prepared to commutate SCR1 on the next half-cycle if that becomes necessary.

11.2.6. Other Circuit Configurations

The two most important additional configurations of the McMurray inverter are shown in Figs. 11-10 and 11-11. Figure 11-10 is the single-phase bridge arrangement. This is essentially two circuits of the type shown in Fig. 11-1. By using two half-bridge circuits, it is not necessary to return the load to a center tap on the dc supply. The opposite leg of the bridge provides the load return path.

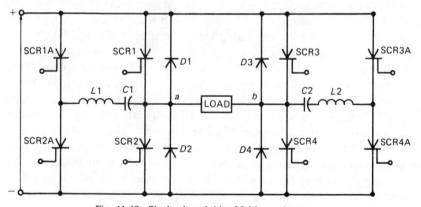

Fig. 11-10. Single-phase bridge McMurray inverter.

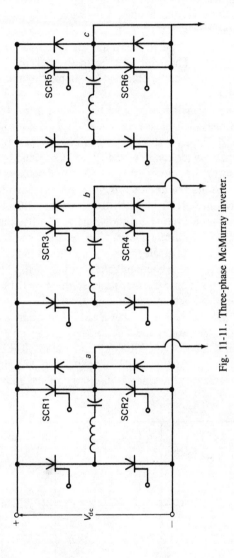

Fig. 11-11. Three-phase McMurray inverter.

Figure 11-11 shows the basic three-phase McMurray inverter. This contains three half-bridge circuits. Thus, the circuit of Fig. 11-1 is the basic building block for both the single-phase bridge and the three-phase McMurray inverter.

The commutating action in Figs. 11-10 and 11-11 is the same as for the circuit of Fig. 11-1. However, the load-voltage waveforms are somewhat more involved. Possibly the simplest way to derive the load-voltage waveforms is to first sketch the waveforms of the voltages at each common point between pairs of complementary SCRs. The load voltage in Fig. 11-10 or the line-to-line voltages in Fig. 11-11 may then be obtained by subtracting the appropriate half-bridge output voltages.

A modified McMurray half-bridge inverter is shown in Fig. 11-12. Diodes $D1A$ and $D2A$ plus resistor R have been added to the circuit in Fig. 11-1. The main purpose of these added components is to limit the maximum commutating capacitor voltage when a wide range of inductive loads must be handled. (Note: The chopper version of this circuit, Fig. 10-14, is discussed in Chapter 10.) For simplicity, to illustrate the circuit principle, an inductive load is assumed. Starting at a time when SCR1A is gated to commutate SCR1, the equivalent circuits are the same as in Fig. 11-9, until time t_3. This is the point at which the commutation capacitor current attempts to reverse, causing SCR1A to go off. If the capacitor voltage is greater than V_{dc}, as shown in Fig. 11-8, there will be an interval of operation with the equivalent circuit of Fig. 11-13. Since first diode $D2$ conducts and then SCR2 conducts, there will be a current flow through the SCR2-$D2$ branch, commutating L and C, resistor R, and diode $D1A$ back to the dc source. Assuming that R is large enough such that this is an overdamped R-L-C resonance, the capacitor voltage will asymptotically approach V_{dc}. When SCR2A is gated to commutate main SCR2, the intervals of operation will resume, as in Fig. 11-9, from t_4 to t_7. At time t_7 there will be another R-L-C overdamped resonance discharge of the commutating capaci-

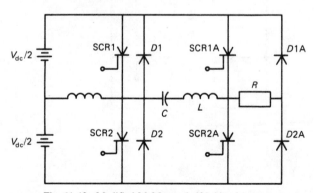

Fig. 11-12. Modified McMurray half-bridge inverter.

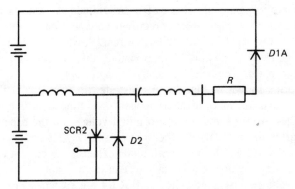

Fig. 11-13. Equivalent circuit for interval between t_2–t_3 and t_3–t_4 in Fig. 11-9.

tor, but now through diode $D2A$. The effect of the addition of diodes $D1A$ and $D2A$ and the resistor R is to limit the commutating capacitor voltage to more nearly equal V_{dc}. This is important because it reduces the voltage ratings required for the SCRs and diodes.

11.3. CURRENT-SOURCE INVERTER

11.3.1. Single-Phase — Resistive Load

A single-phase version of the best known CSI is shown in Fig. 11-14. It is quite similar to the original parallel inverter [3], except that the commutating capaci-

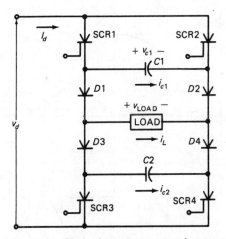

Fig. 11-14. Single-phase current-source inverter.

tor is divided into two parts and diodes $D1-D4$ are added. In a practical application, the circuit is supplied from a rectified voltage with a large reactor in series to provide low ripple in the dc bus current. A constant current source is assumed in Fig. 11-14. This closely approximates the practical situation with a large dc reactor, and it simplifies the analysis appreciably.

The waveforms shown in Fig. 11-15 illustrate the operation with a purely resistive load R and assuming ideal components. At time t_0, it is assumed that SCR1 and SCR4 are triggered with zero initial voltage on the capacitors. The first instant after these ideal switches are closed, $I_d/2$ amperes must flow through each capacitor and through each diode. There is no current through the load resistor R at t_0^+, which is necessary so that there is not an abrupt jump in the capacitor voltages just after SCR1 and SCR4 are switched on. From t_0 to $T/2$, the equivalent circuit is as shown in Fig. 11-16. The capacitor voltages

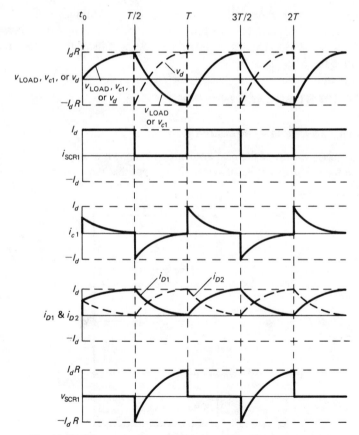

Fig. 11-15. Output waveforms, single-phase CSI with a resistive load.

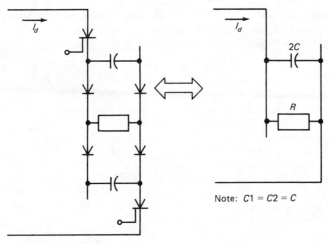

Note: $C1 = C2 = C$

Fig. 11-16. Equivalent inverter circuit during commutation interval.

increase, and this causes current to be diverted from these capacitors to the load resistor. If SCR1 and SCR4 were left closed for a long time, the capacitor voltages would approach $I_d R$, and the current through the load would asymptotically approach I_d. The equations for the capacitor voltage and current from t_0 to $T/2$ may be developed as follows, where for simplicity t_0 is chosen to be zero.

$$I_d = 2C\frac{dv_{C1}}{dt} + \frac{v_{C1}}{R} \tag{11-2}$$

$$\frac{I_d}{s} = 2sCV_{C1}(s) - 2Cv_{C1}(0) + \frac{V_{C1}(s)}{R} \tag{11-3}$$

$$V_{C1}(s) = \frac{I_d + 2sCv_{C1}(0)}{s(2sC + 1/R)} = \frac{(I_d/2C) + sv_{C1}(0)}{s(s + 1/2RC)} \tag{11-4}$$

$$v_{C1}(t) = RI_d(1 - \varepsilon^{-t/2RC}) + v_{C1}(0)\varepsilon^{-t/2RC} \tag{11-5}$$

$$i_{C1}(t) = C\frac{dv_{C1}}{dt} = \frac{I_d}{2}\varepsilon^{-t/2RC} - \frac{v_{C1}(0)}{2R}\varepsilon^{-t/2RC} \tag{11-6}$$

The equations for the next half-cycle of operation may be developed in the same fashion, and similar relations result except that I_d is replaced by negative i_d.

For the waveforms in Fig. 11-15, the RC time constant is assumed relatively short compared with the half-cycle period. As indicated, the initial capacitor voltage when SCR1 and SCR4 are fired is approximately $-RI_d$ volts for the steady-state condition. From Eqs. (11-5) and (11-6), the equations for v_{C1} and i_{C1} in steady state, assuming this initial capacitor voltage, are as follows:

$2n(T/2) < t < (2n + 1)T/2; \quad n = 0, 1, 2 \ldots$

$$v_{C1}(t) = RI_d(1 - 2\varepsilon^{[-(t-nT)/2RC]})$$ (11-7)

$$i_{C1}(t) = I_d \varepsilon^{[-(t-nT)/2RC]}$$ (11-8)

$(2n + 1)T/2 < t < (2n + 2)T/2; \quad n = 0, 1, 2 \ldots$

$$v_{C1}(t) = -RI_d(1 - 2\varepsilon^{-[t-(2n+1)T/2]/2RC})$$ (11-9)

$$i_{C1}(t) = -I_d \varepsilon^{-[t-(2n+1)T/2]/2RC}$$ (11-10)

11.3.2. Single Phase — Inductive Load

For simplicity, a purely inductive load is assumed to illustrate the operation with inductive loads. With SCR1 and SCR4 on, the equations for the capacitor voltages and currents for the case where all diodes are assumed to be conducting are developed as follows. It should be noted that the equivalent circuit is the same as that given in Fig. 11-16, except that the load resistance R is replaced with a load inductance L.

$$I_d = 2C\frac{dv_{C1}}{dt} + \frac{1}{L}\int_0^t v_{C1}\, dt + i_L(0)$$ (11-11)

$$\frac{I_d}{s} = 2sCV_{C1}(s) - 2Cv_{C1}(0) + \frac{1}{sL}V_{C1}(s) + \frac{i_L(0)}{s}$$ (11-12)

$$v_{C1}(s) = \frac{\{[I_d - i_L(0)]/2C\} + sv_{C1}(0)}{s^2 + 1/2LC}$$ (11-13)

$$v_{C1}(t) = [I_d - i_L(0)]\sqrt{\frac{L}{2C}} \sin \omega t + v_{C1}(0) \cos \omega t$$ (11-14)

$$i_{C1}(t) = C\frac{dv_{C1}}{dt} = \frac{I_d - i_L(0)}{2} \cos \omega t - \omega Cv_{C1}(0) \sin \omega t$$ (11-15)

where

$$\omega \triangleq \frac{1}{\sqrt{2LC}}$$

Figure 11-17 illustrates the waveforms, assuming zero initial conditions, ideal components, and that SCR1 and SCR4 are triggered at t_0. With zero initial conditions, the capacitor voltage and current relations are

$$v_{C1}(t) = I_d \sqrt{\frac{L}{2C}} \sin \omega t \qquad (11\text{-}16)$$

$$i_{C1}(t) = \frac{I_d}{2} \cos \omega t \qquad (11\text{-}17)$$

where t_0 is chosen equal to zero to simplify the equations. (Note: In each interval of operation discussed in the remainder of this section, the equations given assume that the beginning of the particular interval occurs at time equal zero.) Figure 11-18 shows the active part of the circuit during each interval of operation.

t_0–t_1. When SCR1 and SCR4 are gated at time t_0, suppose that all of the current I_d were to flow through the branch SCR1-C1-D2-D4-SCR4. This would cause the voltage across C1 to increase while v_{C2} remained at zero. However, a forward voltage would then try to appear across D1, causing it to conduct and thereby connecting the load directly across capacitor C1. Then diode D3 would be forward-biased by the amount that v_{C1} and v_{LOAD} exceeded v_{C2}. A similar argument can be used to show that all of the current, after time t_0, cannot flow through the SCR1-D1-D3-C2-SCR4 branch. Thus, all diodes conduct from time t_0 to t_1, causing a "forced" current sharing by the two capacitors.

When C1 and C2 are in parallel, their voltages must be equal, which means that the capacitor currents must be equal at every instant from t_0–t_1. This is shown in the following manner:

$$v_{C1} = v_{C2} \qquad (11\text{-}18)$$

and thus

$$\frac{dv_{C1}}{dt} = \frac{dv_{C2}}{dt} \qquad (11\text{-}19)$$

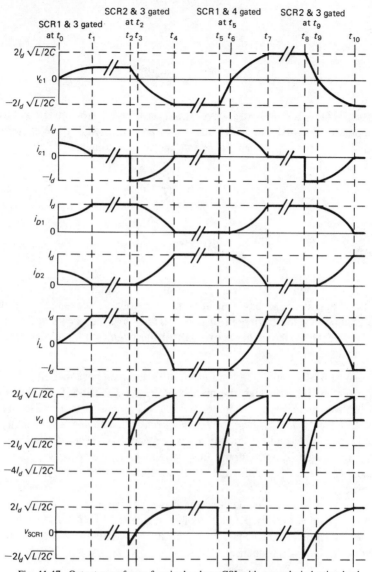

Fig. 11-17. Output waveforms for single-phase CSI with a purely inductive load.

Therefore,

$$i_{C1} = (C1)\frac{dv_{C1}}{dt} = i_{C2} = (C2)\frac{dv_{C2}}{dt} \qquad (11\text{-}20)$$

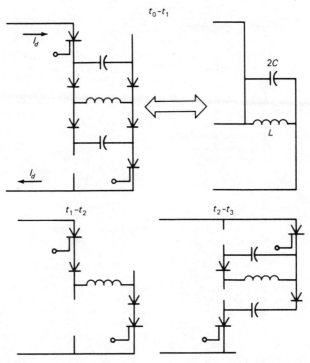

Fig. 11-18. Active circuit during each interval of operation for single-phase CSI with purely induc-tive load.

assuming that $C1$ and $C2$ are equal. Also,

$$i_{D1} + i_{D2} = i_{D1} + i_{C1} = I_d \qquad (11\text{-}21)$$

and

$$i_{D3} + i_{D4} = i_{D4} + i_{C2} = I_d \qquad (11\text{-}22)$$

Since i_{C1} and i_{C2} are equal from Eq. (11-20), i_{D1} and i_{D4} are also equal from Eqs. (11-21) and (11-22). Thus,

$$i_{D2} = i_{C1} = i_{C2} = i_{D3} \qquad (11\text{-}23)$$

and

$$i_{D1} = I_d - i_{C1} = i_{D4} = I_d - i_{C2} \qquad (11\text{-}24)$$

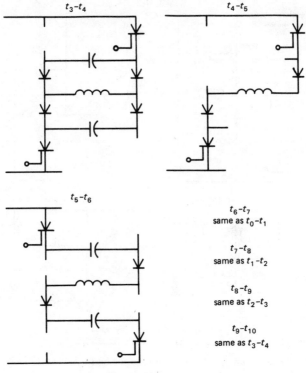

Fig. 11-18. (continued)

The load current is given by the following relation.

$$i_L = i_{D1} - i_{D3} = I_d - 2i_{C1} \tag{11-25}$$

Substituting for i_{C1} from Eq. (11-17) yields

$$i_L = I_d(1 - \cos \omega t) \tag{11-26}$$

$\underline{t_1 - t_2}$.　At time t_1, the capacitor current $i_{C1}(t)$ would attempt to reverse, but this is not possible because of the blocking action of diode $D2$. Similarly, $i_{C2}(t)$ cannot reverse due to $D3$. Thus, from t_1 to t_2, diodes $D2$ and $D3$ block, and the constant current I_d flows through SCR1, $D1$, the load, $D4$, and SCR4.

$\underline{t_2 - t_3}$.　At time t_2, the negative half-cycle of operation is initiated by gating $\overline{\text{SCR2}}$ and SCR3. From t_2 to t_3, the constant current I_d flows through SCR2,

$C1$, $D1$, the load, $D4$, $C2$, and SCR3. This interval ends when v_{C1} and v_{C2} reverse, forward biasing $D2$ and $D3$ respectively.

t_3–t_4. This is similar to the t_0–t_1 interval, except the opposite pair of SCRs is conducting, and there is an initial load current at time t_3 of I_d amperes. The equations for $v_{C1}(t)$ and $i_{C1}(t)$ during this interval are

$$v_{C1}(t) = -2I_d \sqrt{\frac{L}{2C}} \sin \omega t \qquad (11\text{-}27)$$

$$i_{C1}(t) = -I_d \cos \omega t \qquad (11\text{-}28)$$

These relations are derived from Eqs. (11-14) and (11-15), using the conditions at t_3 as initial conditions and replacing I_d with minus I_d wherever it appears. The equations similar to Eqs. (11-21) and (11-22) are

$$i_{D1} + i_{D2} = -i_{C1} + i_{D2} = I_d \qquad (11\text{-}29)$$

$$i_{D3} + i_{D4} = i_{D3} - i_{C2} = I_d \qquad (11\text{-}30)$$

Since i_{C1} and i_{C2} are equal when all four diodes are conducting, as shown from Eqs. (11-18) through (11-20), then from Eqs. (11-29) and (11-30),

$$i_{D1} = -i_{C1} = -i_{C2} = i_{D4} \qquad (11\text{-}31)$$

and

$$i_{D2} = I_d + i_{C1} = I_d + i_{C2} = i_{D3} \qquad (11\text{-}32)$$

Then

$$i_L = i_{D1} - i_{D3} = -i_{C1} - I_d - i_{C1} = -(I_d + 2i_{C1}) \qquad (11\text{-}33)$$

Substituting for i_{C1} from Eq. (11-28),

$$i_L = -I_d(1 - 2 \cos \omega t) \qquad (11\text{-}34)$$

Time t_4 is the point at which diodes $D1$ and $D4$ block when i_{C1} and i_{C2} attempt to reverse.

t_4–t_5. From t_4 to t_5, the constant current I_d flows through SCR2, $D2$, the load, $D3$, and SCR3.

t_5-t_6. At time t_5 the third half-cycle of inverter operation is initiated upon triggering SCR1 and SCR4. At this point, if all diodes began to conduct, the equation for the capacitor current from Eq. (11-15) would be as follows:

$$i_{C1}(t) = I_d \cos \omega t - \omega C v_{C1}(0) \sin \omega t \qquad (11\text{-}35)$$

Since $v_{C1}(0)$ is negative, this would imply that at t_5^+, the current $i_{C1}(t)$ would be greater than I_d. However, this is not possible as it would mean that $i_{D1}(t_5^+)$ was negative. Thus, from t_5 to t_6 the constant current I_d flows through SCR1, C1, D2, the load, D3, C2, and SCR4. This is an interval of operation similar to that during t_2-t_3.

t_6-t_7. At t_6, the capacitor voltages attempt to reverse, but this forward biases D1 and D4. Thus, all diodes conduct, and the equivalent circuit is the same as during t_0-t_1. From Eqs. (11-14) and (11-15), the equations for $v_{C1}(t)$ and $i_{C1}(t)$ are

$$v_{C1}(t) = 2I_d \sqrt{\frac{L}{2C}} \sin \omega t \qquad (11\text{-}36)$$

$$i_{C1}(t) = I_d \cos \omega t \qquad (11\text{-}37)$$

This interval ends when i_{C1} attempts to reverse, which means that D2 and D3 block.

t_7-t_8. From t_7 to t_8, the constant current I_d flows through SCR1, D1, the load, D4, and SCR4.

t_8-t_9. At time t_8, SCR2 and SCR3 are gated to start the fourth half-cycle of inverter operation. The equation for $i_{C1}(t)$, assuming all diodes conduct, is again obtained from Eq. (11-15) with the appropriate initial conditions and replacing I_d with negative I_d.

$$i_{C1}(t) = -I_d \cos \omega t - I_d \sin \omega t \qquad (11\text{-}38)$$

However, this equation would again imply that i_{C1} becomes more negative than $-I_d$ at t_8^+. This is not possible, as it would imply negative current through D2. Thus, from t_8 to t_9 the constant current I_d flows through SCR2, C1, D1, the load, D4, C2, and SCR3.

t_9-t_{10}. At time t_9, the capacitor voltages reach zero, all diodes begin to conduct again, and the equations for $v_{C1}(t)$ and $i_{C1}(t)$ are

$$v_{C1}(t) = -2I_d \sqrt{\frac{L}{2C}} \sin \omega t \qquad (11\text{-}39)$$

$$i_{C1}(t) = -I_d \cos \omega t \qquad (11\text{-}40)$$

When i_{C1} attempts to reverse, t_{10} is reached. From t_{10} until SCR1 and SCR4 are gated again to initiate the fifth half-cycle, the constant current I_d flows through SCR2, $D2$, the load, $D3$, and SCR4. When SCR1 and SCR4 are triggered, the operation is the same as that starting at time t_5. Thus, from this time onward, steady-state operation occurs, and the waveforms are the same on successive cycles of inverter operation.

It should be noted from Fig. 11-17 that the load current is approximately a square wave, assuming $\omega = 1/\sqrt{2LC}$ is considerably smaller than the inverter operating frequency. The wavy lines in Fig. 11-17 are to imply that the relatively long time intervals when I_d is flowing in the load have been compressed so as to show the waveforms more clearly during commutation intervals.

11.3.3. Three-Phase CSI [5]–[9]

The most widely used CSI is the three-phase bridge circuit shown in Fig. 11-19. This also has been called an autosequentially commutated inverter or a six-pulse current-fed inverter with phase sequence commutation. Mittag [5] is credited with the original concept. In certain adjustable-speed ac motor drive applications, this CSI can have the following advantages.

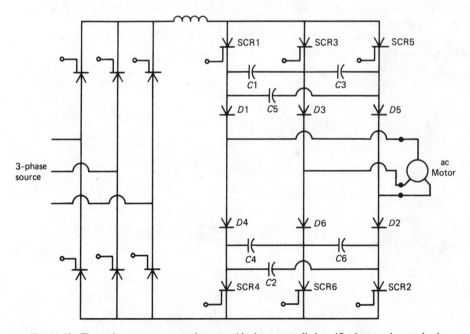

Fig. 11-19.   Three-phase current-source inverter with phase-controlled rectifier input and motor load.

1. Fast turn-off thyristors are not required.
2. The output short-circuit current is inherently limited to the inverter dc input-current level.
3. When the dc input current is fed from a phase-controlled rectifier operating from an ac supply, regenerative electrical braking is obtained easily by operating the phase-controlled rectifier in its inverting mode, since the dc voltage v_d can have a negative average value for this CSI.

A complete analysis of the circuit in Fig. 11-19 is quite complex, since there are multiple modes of operation, some of which are unstable. Thus, the most practical means of predicting the behavior of this inverter is from a digital computer simulation. A general discussion of one operating mode is presented here to form the basis for the development of a simulation program necessary to perform a detailed computer-aided analysis of this three-phase CSI.

For simplicity, the inverter input is considered to be an ideal constant current source I_d. The SCR gating signals are assumed to be as shown in Fig. 11-20. Each SCR is gated continuously for 120°, and there is 60° phase displacement between the gating signals for the odd- and even-numbered devices. The first interval of operation begins at time t_0, the moment of gating SCR1, which instantly commutates SCR5 because of the initial voltages on capacitors $C1$, $C3$, and $C5$. Prior to time t_0, SCR5 is assumed to have been conducting the constant current I_d into line 3 of the motor, which then flowed out of line 2 through SCR6 and back to the current source. The equivalent circuit for interval t_0-t_1 is shown in Fig. 11-21(a). This is a constant-current charging interval for capacitors $C1$, $C3$, and $C5$. Assuming the capacitors are equal.

$$C_1 = C_2 = C_3 = C_4 = C_5 = C_6 = C \qquad (11\text{-}41)$$

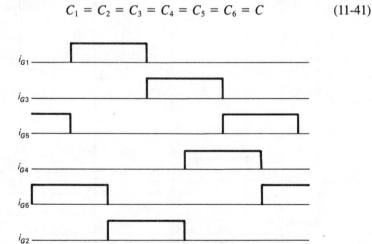

Fig. 11-20. SCR gating pattern.

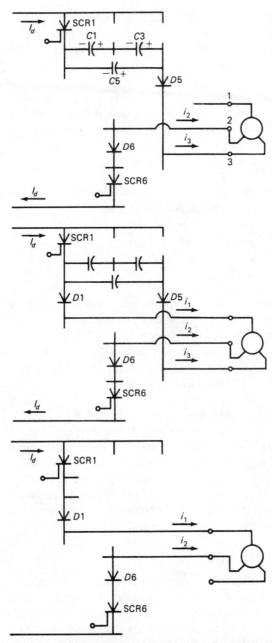

Fig. 11-21. Equivalent circuits. (a) Equivalent circuit during t_0–t_1 interval. (Note: The positive polarity for each capacitor voltage is indicated.) (b) equivalent circuit during t_1–t_2 interval; (c) equivalent circuit during t_2–t_3 interval.

and the series combination of $C1$ and $C3$ results in $C/2$ in parallel with the C due to $C5$. Thus, the current I_d divides into $I_d/3$ through $C1-C3$ and $2I_d/3$ through $C5$. During this interval the motor line currents are constant:

$$i_3 = I_d; \qquad i_2 = -I_d \qquad (11\text{-}42)$$

The motor line-to-line voltages are established by the motor parameters and operating conditions, including the shaft load. Interval t_0-t_1 ends when

$$v_{3-1} = v_{C5} \qquad (11\text{-}43)$$

The equivalent circuit for the t_1-t_2 interval is shown in Fig. 11-21(b). During this interval, the important circuit equations are

$$v_{C5} = v_{C1} + v_{C3} = v_{3-2} - v_{1-2} = v_{3-1} \qquad (11\text{-}44)$$

$$i_1 + i_3 = I_d = -i_2 \qquad (11\text{-}45)$$

Again, the motor parameters and operating conditions determine the motor voltages. The t_1-t_2 interval ends when the motor line current i_3 falls to zero and attempts to reverse.

The equivalent circuit during interval t_2-t_3 is shown in Fig. 11-21(c). The constant current I_d flows through SCR1, $D1$, into motor line 1, out of motor line 2, and through $D6$ and SCR6 back to the source. Again the motor voltage is determined by the motor parameters and operating conditions.

Interval t_2-t_3 ends 60° after time t_0 when SCR2 is gated. This initiates another sequence of three intervals similar to those of Fig. 11-21(a)–(c). In fact, three similar intervals occur when each of the remaining SCRs are gated. Thus, there are 18 intervals of operation during each complete cycle.

This brief discussion was intended to describe one mode of operation for the circuit of Fig. 11-19. As indicated previously, there are additional operating modes, depending upon the circuit parameters, operating frequency, and the particular machine used to drive a given load [8].

11.4. ADDITIONAL REMARKS

In this chapter, the operation of two important forced commutated inverters is discussed. This is only an introduction to this inverter family; there are numerous additional schemes which have been developed and used commercially. Also, high-power transistor and GTO inverters are finding increased application in situations where the McMurray inverter or the CSI might have been used previously.

REFERENCES

1. McMurray, W. SCR Inverter Commutated by an Auxiliary Impulse. IEEE Transactions on Communications and Electronics 83, no. 75: 1964, pp. 824–829.
2. Bedford, B. D., and Hoft, R. G. *Principles of Inverter Circuits.* New York: John Wiley & Sons, 1964, chap. 7.
3. Ibid., chap. 4.
4. Pollack, Jerry J. Advanced Pulsewidth Modulated Inverter Techniques. IEEE Transactions on Industry Applications, IA-8, no. 75: March/April 1972, pp. 145–154.
5. Mittag, A. H. Electric Valve Converting Apparatus. U.S. Patent 1,946,292 (to General Electric), Feb. 6, 1934.
6. Ward, E. E. Inverter Suitable for Operation over a Range of Frequency. Proc. IEE 111, no. 8: August 1964, pp. 1423–1434.
7. Bose, Bimal K., ed. *Adjustable Speed AC Drive Systems.* New York: IEEE Press, 1981, pp. 193–268.
8. Lienau, W. Commutation Modes of a Current Source Inverter. Proc. of 2nd IFAC Symposium on Control in Power Electronics and Electrical Drives: Pergamon Press, 1977, pp. 219–229.
9. Moltgen, Gottfried. *Converter Engineering: An Introduction to Operation and Theory.* Berlin: Siemens Aktiengesellschaft and John Wiley & Sons, Ltd., 1984, pp. 132–143.

PROBLEMS

1. Calculate the commutating inductance L and capacitance C required for the McMurray half-bridge inverter of Fig. 11-1 assuming the following:
   a) The voltage across the commutating capacitor just prior to commutation is 500 V.
   b) The load current is essentially constant at 100 A during commutation.
   c) There is negligible loss in the commutating circuit.
   d) A turn-off time of 40 μs is required for the main thyristors.
   e) The peak commutating current $I_{C, \text{peak}}$ is 200 A.

2. Consider the circuit of Fig. 11-1 with $V_{\text{dc}} = 250$ V, a resistive load $R = 1$, the commutating $L = 32$ μH, the commutating $C = 8$ μF, a commutating capacitor voltage of 500 V just prior to SCR1A or SCR2A gating, and an inverter operating frequency of 1000 Hz. Determine the approximate RMS current ratings required for SCR1, SCR2, SCR1A, and SCR2A.

3. Suppose that the McMurray half-bridge inverter of Fig. 11-1 is to be applied in a case where a purely resistive or a purely inductive load may occur, with the same peak commutating current required for either type of load.
   a) What load condition determines the earliest moment at which the next main SCR is gated after gating an auxiliary SCR to initiate commutation?
   b) Explain the major disadvantages of maintaining the same time interval between auxiliary SCR gating and the instant of gating the next main SCR, when the load varies over the complete range of current magnitude and power factor.

4. Calculate the value of $C = C1 = C2$ necessary for the single-phase CSI circuit of Fig. 11-14 assuming the following:
   a) Resistive load with $R = 2.5\ \Omega$.
   b) An RC time constant which is short relative to the inverter half-period.
   c) The dc current $I_d = 100$ A.
   d) A 50-μs turn-off time required by the SCRs.
5. Consider the circuit of Fig. 11-19 with an ideal synchronous motor load. Assume that the ideal motor may be represented by a three-phase wye-connected set of sinusoidal voltage sources with a series resistance R and inductance L in each motor line. Assuming the three operating modes discussed in Section 11.3.3, write all of the equations necessary during each of the three operating modes for a digital computer simulation. Explain the circuit operating conditions which cause the change in circuit topology at the end of each interval throughout one complete cycle.

APPENDIX I
SEMICONDUCTOR PARAMETERS

2 SELECTING THE PROPER POWER SEMICONDUCTOR FOR YOUR APPLICATION

1. BASIC SEMICONDUCTOR PARAMETERS DEFINED

The primary tool for selecting the proper semiconductor to meet a given requirement is the technical data sheet. In order to use the data sheets effectively, one must first understand the significance of some of the more important parameters used in rating these semiconductors. The following apply to all types of semiconductors:

$T_{J(MAX)}$—Maximum allowable junction temperature. This is a very critical parameter upon which all device ratings are based. Typically, most SCR's have an upper operating junction temperature limit of 125°C (with some high temperature series up to 150°C capability); most transistors have an upper operating junction temperature limit in the range of 150°C to 175°C; and most rectifiers have an upper operating junction temperature limit in the range of 175°C to 200°C. If these respective values are exceeded, the device becomes vulnerable to rapid failure. Of course, proper derating below these maximum limits is necessary in the actual application to assure reliable performance.

$T_{c(MAX)}$—Maximum allowable case temperature. Since the user of semiconductors has no way of actually measuring the junction temperature inside of a device package, the case temperature becomes a practical, useful value which can be measured and monitored. As a result, data sheets usually show the current carrying capability of a device as a function of its case temperature. The case temperature is directly related to the junction temperature by the device characteristic, $R_{\theta JC}$ (thermal impedance).

$R_{\theta JC}$—Thermal impedance, junction to case. $R_{\theta JC}$ is measured in degrees centigrade per watt and is determined by the device construction and materials used. This parameter indicates the ability of the device to transfer heat away from the junction and out through the device case—the lower the value, the better.

Other basic device parameters by family type that are essential for the semiconductor user to know in the selection effort are as follows:

Family Type	Input/ Control	Output	Losses	Overload	Other Special
RECTIFIERS	NONE	$I_{F (av)}^{(1)}$ $V_{RRM}^{(2)}$	$V_{FM}^{(3)}$, $I_{RRM}^{(4)}$	$I_{FSM}^{(5)}$	$t_{rr}^{(6)}$
SCR's	$I_{GT}^{(6)}$	$I_{T(av)}^{(1)}$, $V_{RRM}^{(2)}$	$V_{TM}^{(3)}$, $I_{RRM}^{(4)}$	$I_{TSM}^{(5)}$	di/dt,$^{(7)}$ dv/dt,$^{(8)}$ t_{on},$^{(9)}$ t_q $^{(10)}$
TRANSISTORS	$I_B^{(6)}$	I_C,$^{(1)}$ $V_{CE}^{(2)}$	$V_{CE (SAT)}$,$^{(3)}$ $V_{BE (SAT)}$,$^{(4)}$ $I_{CEO}^{(5)}$	NONE	SOA,$^{(7)}$ t_{on},$^{(8)}$ t_{off},$^{(9)}$ h_{FE} $^{(10)}$

Table 2.1 BASIC SEMICONDUCTOR DEVICE PARAMETERS

The following expansion of the items in the table explain their importance:

RECTIFIERS

(1) $I_{F(av)}$—Maximum full cycle average forward current (measured with an average reading meter) at a

specified case temperature. This is the maximum amount of current that can be controlled or rectified by any given device.

(2) V_{RRM} —Maximum repetitive peak reverse voltage. This is the maximum allowable voltage that a device will block in the reverse direction. An associated parameter, V_{RSM}, defines the device capability to withstand non-repetitive reverse voltage peaks of less than five millisecond duration. This parameter typically ranges from zero to 25% above the repetitive values.

(3) V_{FM} —Maximum forward voltage drop at a specified forward current and device case temperature. The large majority of the losses in a rectifier are due to the forward voltage drop. Therefore, low forward drops are very desirable in high power devices because less power is dissipated ($P=V_{FM} \times I_{F(av)}$).

(4) I_{RRM} —Maximum reverse leakage current at V_{RRM}. Low reverse leakage currents are desirable for the same reason as a low V_{FM}, less power (watts) to be dissipated ($P=V_{RRM} \times I_{RRM}$).

(5) I_{FSM} —Maximum one-half cycle ($60H_Z$) peak surge current (under load). Ability to withstand current surges in excess of the rated operating current extends the application capability of a rectifier. This characteristic is non-repetitive in nature and, by definition, may occur only 100 times within the life of the device. An associated parameter, I^2t, is used to coordinate sub-cycle fuse clearing time with sub-cycle fault rating of a rectifier.

(6) t_{rr} —Maximum reverse recovery time relates to the time required for the reverse current spike to dissipate after a rectifier stops conducting forward current. This characteristic is especially important on a class of rectifiers called "fast recovery" devices. Because all rectifiers are essentially "self-operated" switches, they have finite switching times. This characteristic becomes important when one wishes to rectify (or switch) at high frequencies.

SCR's

(1) $I_{T(av)}$ —Maximum allowable average forward current at a specified conduction angle and case temperature. If a conduction angle of 180° is specified, this parameter would be the same as $I_{F(av)}$ in a rectifier. However, unlike a rectifier, the conduction angle of an SCR can be controlled from 0 to 180° (typically, a data sheet lists five or six of these angles). The conduction angle defines that part of the sine wave during which the device is conducting current.

(2) V_{RRM} —Maximum repetitive reverse blocking voltage. The same comments apply here as for V_{RRM} of a rectifier. The SCR in addition to having reverse blocking capability will also block voltage in the forward direction. V_{DRM} is the designation for the maximum value of repetitive forward blocking voltage. Most SCR designs are symmetrical in nature—that is, the forward and reverse repetitive blocking voltage capabilities for a given device are equal ($V_{DRM} = V_{RRM}$). Both repetitive voltage ratings (V_{RRM} V_{DRM}) have corresponding non-repetitive voltage ratings, V_{RSM} and V_{DSM}.

(3) V_{TM} —Maximum forward voltage drop at a specified forward current and device case temperature. The same comments apply here as for V_{FM} in a rectifier.

(4) I_{RRM} —Maximum reverse leakage current at V_{RRM}. Again, as with a rectifier, low leakage currents, I_{RRM}, are desirable as less power will be dissipated in the blocking (off-state) mode. In addition, as an SCR has repetitive forward voltage capability, V_{DRM}, it also has a forward leakage current, I_{DRM}.

(5) I_{TSM} —Maximum peak surge current for a given number of cycles at $60H_Z$. As with a rectifier, this characteristic is by definition non-repetitive and may occur only 100 times within the life of the device. In addition, following this current surge, the repetitive forward blocking voltage, V_{DRM}, is not guaranteed. This subject is treated in more detail in the ⓦ Thyristor Surge Suppression Ratings Application Data Sheet in the General section of the Data Book. An associated parameter, I^2t, is used for fuse coordination.

(6) I_{GT} —Minimum DC gate current to trigger (turn-on) an SCR at stated conditions of temperature and forward blocking voltage. This characteristic specifies the absolute minimum amount of current that must be provided from a logic source to switch an SCR from blocking voltage (off-state) to conducting current (on-state). For more information, see the SCR Gate Turn-on Characteristics Application Data Sheet in the General section of the Data Book.

(7) di/dt—Maximum rate of rise of peak current allowable with respect to time during SCR turn-on. Because an SCR requires a finite time to turn on, only a very small area of the junction is conducting current at the instant the device is triggered on. If the current is building up very quickly, it must all go through this very small area. When the specified di/dt limit is exceeded, the device can develop a hot spot which can destroy the unit. This is generally not a problem in $60H_Z$ phase control applications but becomes a problem in DC switching applications (e.g. inverters) or in capacitor discharge circuits. When comparing di/dt ratings of different manufacturers be sure the ratings are both repetitive or nonrepetitive and that the ratings are based on the same set of test conditions.

(8) dv/dt—Minimum rate of rise of peak voltage with respect to time which will cause switching from the off-state to the on-state. When the dv/dt limit is exceeded, the potential danger exists of turning the SCR back on when it should remain off. As with di/dt this is generally only a problem in square wave or step function applications. Review all test conditions when comparing dv/dt ratings of different manufacturers—dv/dt can be expressed as either a linear or an exponential function.

(9) t_{on} —Time required for the forward current to reach 90 percent of its final (or maximum) value when switching from the off-state to the on-state under specified conditions. Switching times are important for two reasons—they affect the upper frequency at which the device may be operated and to some degree they determine the system efficiency. When considering frequency of operation (or in pulse applications, the desired pulse width), it becomes simply a matter of whether or not the device can be turned on and off fast enough to satisfy the requirements. Power losses when switching from blocking voltage to conducting current can be a significant consideration for determining system efficiency.

(10) t_q —Turn-off time relates to the time required for an SCR to switch from conducting current to blocking forward voltage. Turn-off time is important for essentially the same reasons as turn-on time. Turn-off time is not critical in most 60 cycle phase control applications; however, a special class of SCR's called "fast switching" devices are used in inverters, choppers, and other high frequency circuits. In these types of applications, the designer must be careful to select a device offering the optimum combination of current handling capability, voltage blocking capability, and turn-off time capability. Due to the wide variety of fast switching applications, data sheet turn-off conditions do not necessarily reflect actual circuit operating conditions. Contact ⓦ if help is needed in selecting the proper device.

TRANSISTORS

(1) I_C —Collector current. In most applications, this is the current that is manipulated to perform some desired function. In a series regulator, for example, the collector current can be increased or decreased by base current control, depending upon what the load requires.

(2) V_{CE} —Collector-to-emitter voltage. This is similar to V_{RRM} in rectifiers and SCR's in that it represents the device blocking capability in the off-state. There are many related designations: V_{CEO} , V_{CES} , V_{CER} , etc. They differ only in the third subscription letter which indicates the condition of the base: O-open, R-resistor, S-shorted, etc.

(3) $V_{CE(SAT)}$ —Collector-emitter saturation voltage. When driving the base of a transistor, a point is reached where increased base current no longer results in decreased collector-emitter voltage. This is saturation. $V_{CE(SAT)}$ is a measure of the voltage across that junction under saturation and is comparable to forward voltage drop in rectifiers and SCR's. $V_{CE(SAT)}$ and I_C lead to power losses in the transistor during saturated operation and therefore, are important considerations.

(4) $V_{BE(SAT)}$ —Base-emitter saturation voltage. The same general comments apply here as for $V_{CE(SAT)}$ except the comments now refer to a base-emitter rather than a collector-emitter condition. High $V_{BE(SAT)}$ or large variations in $V_{BE(SAT)}$ will cause corresponding changes in $V_{CE(SAT)}$

(5) I_{CEO} —Collector-to-emitter leakage current with the base open. This is usually the primary leakage loss in a transistor. This characteristic is comparable to I_{RRM} in rectifiers or SCR's. While low values of leakage are desirable to minimize power losses, low leakage is not necessarily synonymous with reliability.

(6) I_B —Base current. The function of the base current in a transistor is similar to that of the gate current in an SCR. However, in a transistor, current must be provided into the base as long as the transistor is to be kept on, whereas, the SCR only requires an initial pulse of current to turn it on.

(7) SOA—Safe Operating Area. SOA is a voltage-current plot which describes an area in which the transistor can operate safely. A time limit is given for the collector voltage and collector current that can occur simultaneously in the transistor. Forward bias SOA ratings require the base-emitter to be forward-biased throughout the time that the peak power condition exists and is usually measured in a resistive circuit. A related term is forward current stability, denoted by the symbol I_{SB} . Safe operation of a transistor during inductive switching when a transistor in series with an inductor is turning off necessitates consideration of additional characteristics. The inductance will keep current flowing for some period of time. During this time, the voltage is increasing across the transistor creating a VxI product or power dissipation. Because the base-emitter is reverse-biased, the transistor cannot dissipate as much power. The energy that a transistor can support is denoted by the abbreviation E_{SB} . Inductive switching ratings can be used as a guide to compare transistor capabilities, but actual capabilities must be verified in the actual circuit.

(8) & (9) t_{on} & t_{off} —Turn-on time and turn-off time. These are important parameters in transistors for essentially the same reasons given in the SCR remarks.

(10) h_{FE} —DC current gain under specified conditions of collector current and collector-emitter voltage. This transistor characteristic is the ratio of DC collector current to DC base current. This amplification factor determines the amount of output (I_C) that is generated by a given input (I_B). For example, if h_{FE} is 20 at I_B =1 ampere, then I_C is 20 amperes.

2. CONFUSING TERMINOLOGY—MANUFACTURER VERSUS USER INTERPRETATION

Many users and designers have either been lucky or have learned the hard way how to properly interpret a semiconductor manufacturer's data sheet. Unfortunately, many conventional semiconductor terms and definitions can have double meanings, depending upon whether they are being interpreted from the manufacturer's or the user/designer's point of view. What a semiconductor manufacturer might define as a maximum (minimum) value on a data sheet could well be a minimum (maximum) value for the designer and user. A few examples utilizing SCR terminology will illustrate this dilemma:

Table 2.2 Confusing Data Sheet Terminology

Data Sheet Terminology	User or Designer's Interpretation Might Be	Semiconductor Mfgr's Actual Meaning	Remarks
(1) Maximum $I_{GT}=150$ ma	Does this mean that no more than 150 ma is needed to turn-on these SCR's?	All SCR's supplied will have gate currents less than or equal to 150 ma. User must supply *more than* 150 ma to assure proper SCR turn-on. In fact, 3 to 5 times as Max. I_{GT} is desirable for SCR turn-on in certain applications.	Here, a value defined as maximum by the manufacturer is for the user or designer, an absolute minimum design limit.
(2) Maximum blocking or off-state voltage V_{DRM}, V_{RRM} =1200 volts.	Does this mean that some devices received will block or support less than 1200 volts?	All devices supplied will support *at least* 1200 volts; however, this limit *cannot* be exceeded in the application.	In this case, the manufacturer's maximum is also, the designer's or user's maximum limit.
(3) Minimum di/dt – 200 A/μs	Does this mean that 200 A/μs can be exceeded in a given application?	All devices supplied will withstand a rate of current rise (di/dt) of *at least* 200 A/μs; however, this limit *cannot* be exceeded in the application.	Here, the manufacturer's minimum value becomes the user's or designer's maximum design limit.
(4) Typical turn off time $t_q = 40\mu$s	Can this typical turn-off time be relied on in a given application?	This 40μs turn-off time *only* represents an average value for the product family and is *not* a guaranteed limit.	Typical values should *not* be relied upon by the designer or user as guaranteed values.

APPENDIX II
D62T TRANSISTOR

NPN Power Switching
TRANSISTORS
D62T

200 Amperes
400—500 Volts

| Symbol | Inches | | Millimeters | |
	Min.	Max.	Min.	Max.
ϕD	1.610	1.650	40.89	41.91
ϕD_1	.745	.755	18.92	19.18
ϕD_2	1.420	1.460	36.07	37.08
H	.500	.560	12.70	14.22
ϕJ	.135	.145	3.43	3.68
J_1	.072	.082	1.83	2.08
L	4.000		101.6	
N	.030		.76	

Creep Distance—.34 in. min. (8.64mm)
Strike Distance—.52 in. min. (13.21mm)
(In accordance with NEMA standards.)
Finish—Nickel Plate.
Approx. Weight—2.1 oz. (60 g).
1. Dimension "H" is a clamped dimension.

2. "Base Lead is No. 14 uninsulated
flexible stranded wire.

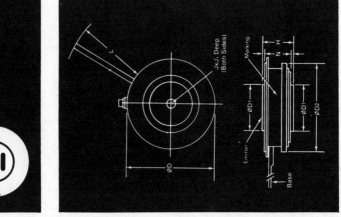

D62 Outline

Maximum Ratings
Collector Current (peak): 200 Amperes
Collector Current (continuous): 200 Amperes
Base Current (continuous): 20 Amperes
Power Dissipation: 1100 Watts at $T_C = 75^\circ C$
Operating and Storage Temperature: -65°C to 200°C

Applications
● High Frequency Inverters
● Motor Controls
● Switching Regulators
● VLF Transmitters
● Induction Heating
  Power Supplies

Features
● Triple Diffused Design
● CBE Construction
● Double Sided Cooling
● Fast Switching

Ordering Information

Type	V_{CEO} (sus) (Volts)	Current Rating — Amperes				Gain
		40	50	60		
D62T	400	4040	4050	4060		10
	450	4540	4550			10
	500	5040				10

Example: Select the complete ten digit device part number you desire from the shaded area in the table above — i.e. a D62T454010 Describes a Disc Package Transistor rated at 450 Volts, 40 Amperes, and a gain of 10 at rated current (40 Amperes).

Electrical and Mechanical Characteristics (TCASE = 25°C unless otherwise specified)

Symbol	Characteristic	Test Conditions	Min.	Typ.	Max.	Units
VCEO (SUS)	Collector-Emitter Sustaining Voltage	IC = 200mA IB = 0 Note 1		See Ordering Information on Previous Page		Volts
ICEV	Collector Cutoff Current (Base Emitter Reverse Biased)	At Rated VCEO (SUS) · 50V VBE (OFF) = -1.5V.		0.1	1	mA
ICEV	Collector Cutoff Current (Base Emitter Reverse Biased)	At Rated VCEO (SUS) · 50V VBE (OFF) = -1.5V. TC = 150°C		0.8	3	mA
IEBO	Emitter Cutoff Current	VEB = 7V		2	30	mA
hFE	DC Current Gain	IC · Gain Rated · VCF 2.5V	10	15		
hFE	DC Current Gain D62TXX40	IC = 80A, VCE = 2.5V		5		
hFE	DC Current Gain D62T—-50	IC = 10uA , VCE 2.5V		5		
hFE	DC Current Gain D62T · 60	IC =120A, VCE = 2.5V		5		
VCE (SAT)	Collector-Emitter Saturation Voltage	IC = Gain Rated IC/IB = 8.33		75	1.25	Volts
VBE (SAT)	Base-Emitter Saturation Voltage	IC/IB = 8.33 D62T_40 D62T_50 D62T_60		1.0 1.15 1.30	1.40 1.50 1.60	Volts

	Parameter	Conditions	Min	Typ	Max	Units
td	Turn-On Delay	Vcc = 250V, Ic = 40A			110	ns
tr	Rise Time	Resistive Load Switch Times — IB1 = IB2 = 6A		0.7	1.0	µs
ts	Storage Time	tp = 50 us		1.25	3.0	µs
tf	Fall Time	Duty Cycle ≤ 2% D62T—40		0.3	0.5	µs
td	Turn-On Delay	Vcc = 250V, Ic = 50A			120	ns
tr	Rise Time	Resistive Load Switch Times — IB1 = IB2 = 7.5A		0.8	1.10	µs
ts	Storage Time	tp = 50 us		1.75	3.0	µs
tf	Fall Time	Duty Cycle ≤ 2% D62T—50		0.32	0.5	µs
td	Turn-On Delay	Vcc = 250V, Ic = 60A			150	ns
tr	Rise Time	Resistive Load Switch Times — IB1 = IB2 = 9A		0.85	1.20	µs
ts	Storage Time	tp = 50 us		2.0	3.0	µs
tf	Fall Time	Duty Cycle ≤ 2% D62T—60		0.35	0.5	µs
COB	Output Capacitance	fTEST = 1 MHz, VCB = 10V		2500		µµf
fT	Gain-Bandwidth Product	fTEST = 1 MHz, IC = 5A, VCE = 10V	7	10		MHz
RθJC	Thermal Resistance Junction to Case Double Sided Cooling	VCE = 20V			0.09	°C/W
RθCS	Thermal Resistance Case to Sink Double Sided Cooling	VCE = 20V Lubricated			0.05	°C/W
	Mounting Force		900		1100	lb.
			4.05		4.95	KN

1. VCEO (SUS) must not be measured on a curve tracer.

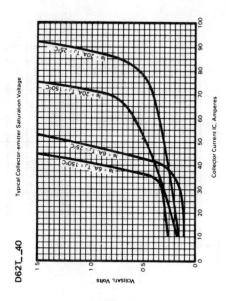

D62T__40

Typical Collector-emitter Saturation Voltage

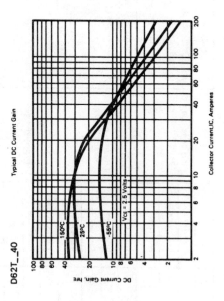

D62T__40

Typical DC Current Gain

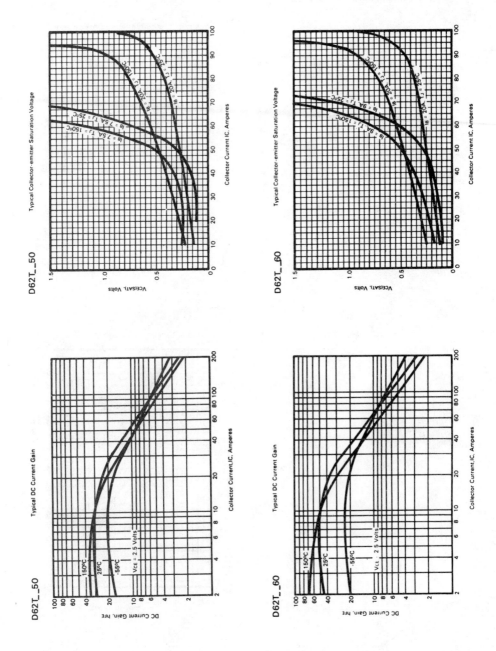

D62T_50 — Typical Collector-emitter Saturation Voltage

D62T_60 — Typical Collector-emitter Saturation Voltage

D62T_50 — Typical DC Current Gain

D62T_60 — Typical DC Current Gain

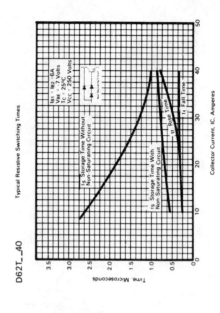

Typical Resistive Switching Times

D62T_40

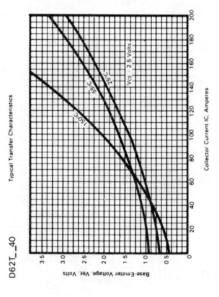

Typical Transfer Characteristics

D62T_40

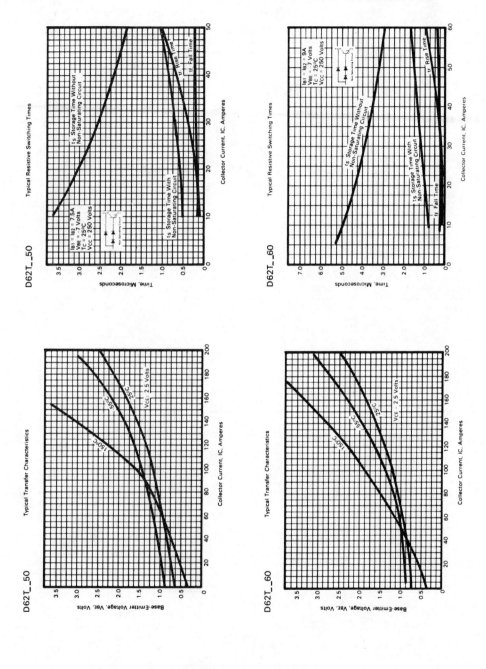

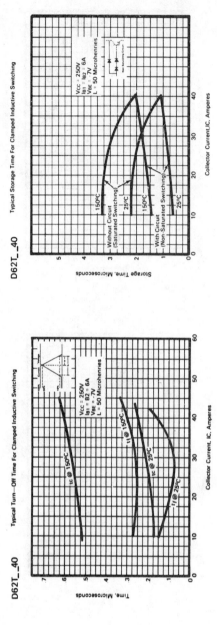

D62T_40   Typical Storage Time For Clamped Inductive Switching

D62T_40   Typical Turn—Off Time For Clamped Inductive Switching

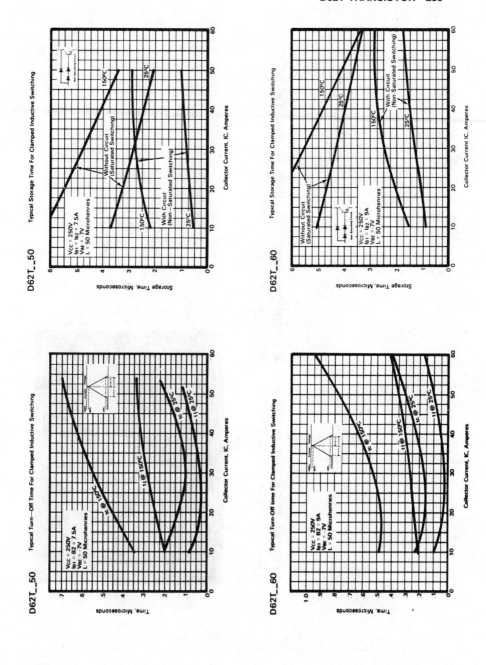

D62T_50   Typical Storage Time For Clamped Inductive Switching

D62T_60   Typical Storage Time For Clamped Inductive Switching

D62T_50   Typical Turn—Off Time For Clamped Inductive Switching

D62T_60   Typical Turn—Off time For Clamped Inductive Switching

TYPICAL TURN-OFF WAVE FORMS FOR CLAMPED INDUCTIVE SWITCHING*

* Shown below are actual photographs taken during 150°C inductive switching measurements

D62T_40

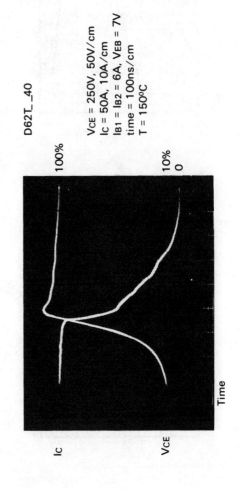

V_{CE} = 250V, 50V/cm
I_C = 50A, 10A/cm
I_{B1} = I_{B2} = 6A, V_{EB} = 7V
time = 100ns/cm
T = 150°C

100%

10%

0

Ic

V_{CE}

Time

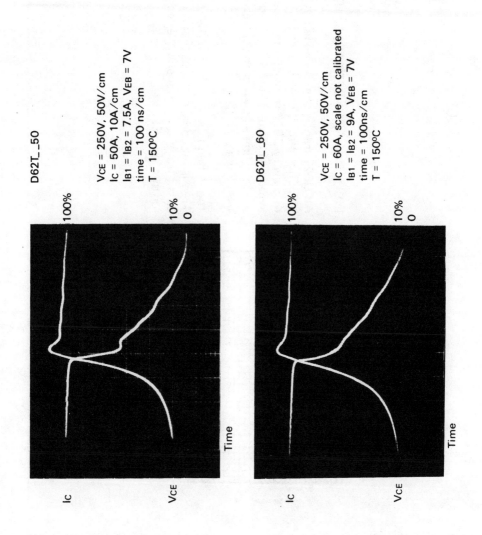

D62T_50

V_{CE} = 250V, 50V/cm
I_C = 50A, 10A/cm
I_{B1} = I_{B2} = 7.5A, V_{EB} = 7V
time = 100 ns/cm
T = 150°C

100%

10%
0

I_C

V_{CE}

Time

D62T_60

V_{CE} = 250V, 50V/cm
I_C = 60A, scale not calibrated
I_{B1} = I_{B2} = 9A, V_{EB} = 7V
time = 100ns/cm
T = 150°C

100%

10%
0

I_C

V_{CE}

Time

INDUCTIVE TURN-OFF CAPABILITY

All D62T transistors are tested in the clamped inductive circuit shown. This test is assurance that every D62T transistor is capable of switching clamped inductive loads traversing the load line as shown on the Reverse Bias Safe Switching graph.

The dotted line on the Safe Switching Area is an additional test that can be done on all transistors. Consult the factory for this special test.

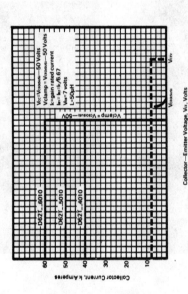

Reverse Bias Safe Switching Area

$V_{CC}=V_{CE(sus)}$—50 Volts
$V_{Clamp}>V_{CE(sus)}$—50 Volts
I_C=gain rated current
$I_B=-I_C/6.67$
$V_{EB}=7$ volts
$L=50\mu H$

$V_{Clamp}=V_{CE(sus)}$—50V

Collector Current, I_C Amperes

Collector—Emitter Voltage, V_{CE} Volts

FORMULA TO DETERMINE SWITCH INTERVAL JUNCTION TEMPERATURE EXCURSIONS*

$$T_{max} = \left[\frac{t_p \, Z\theta_1}{T} + \left[1 - \frac{t_p}{T} \right] Z\theta_2 - Z\theta_3 + Z\theta_4 \right] P + T_C$$

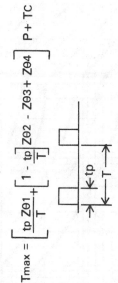

$Z\theta_1 = Z\theta_{JC}$ Steady State

$Z\theta_2 =$ Transient Thermal Impedance of Time $(T + t_p)$

$Z\theta_3 =$ Transient Thermal Impedance of the Period T

$Z\theta_4 =$ Transient Thermal Impedance of the Pulse Width t_p

$P =$ Peak Power During Switching

$T_{max} =$ Maximum Junction Temperature

$T_C =$ Case Temperature

*For further information consult factory.

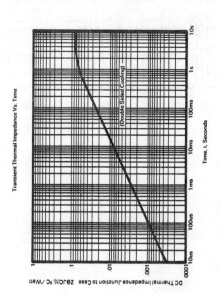

Transient Thermal Impedance Vs. Time

(Double Sided Cooling)

Time, t, Seconds

DC Thermal Impedance Junction to Case $Z\theta JC(t)$, °C /Watt

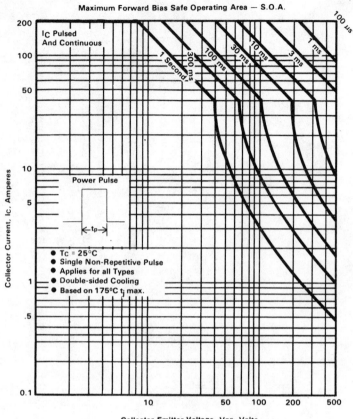

Maximum Forward Bias Safe Operating Area — S.O.A.

APPENDIX III
APPLICATION DATA—POWER SWITCHING
TRANSISTOR D60T

Transistors are capable of controlling large amounts of power at high frequencies but with ever increasing power requirements, paralleling transistors is required. The Westinghouse D60T transistor is a high current, high voltage transistor that eliminates paralleling in many applications. The D60T

(Not to scale)

D60T
Massive compression bonded copper emitter contact

Competitor A (Modified TO-3)
Small diameter, point contact wire bonds

Competitor B (TO-63)
thin copper emitter solder contact

Westinghouse's D60T compression bonded encapsulation (CBE) construction with a massive emitter rod contact provides:
- High current overload capability (no bonding wire current limit)
- Better power cycling capability (no solder fatigue)

Figure 1: Comparison of Power Transistor Contacting Methods

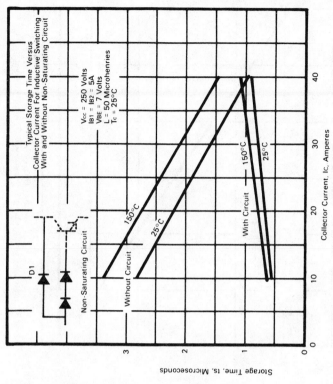

Figure 2: Non-Saturating Circuit Performance

Typical Storage Time Versus Collector Current For Inductive Switching With and Without Non-Saturating Circuit

$V_{CC} = 250$ Volts
$I_{B1} = I_{B2} = 5A$
$V_{BE} = 7$ Volts
$L = 50$ Microhenries
$T_C = 25°C$

uses new and innovative packaging that reduces the power dissipation limitations of high frequency, high voltage, high current transistors. This application data sheet discusses the limitations of power transistors and also presents many common circuit configurations.

Current Capability

Most transistors use ultrasonically bonded wire or soldered clip leads to connect the silicon emitter area to the package emitter pin. These wires are necessarily small in diameter limiting the maximum current of the transistor. Also, ultrasonically bonding the wires to the emitter limits the contact to a small area. This "bonding wire" limits not only the continuous current but also limits the maximum average current during duty cycle use. An example of current limitations is when a pulse width modulated circuit operates satisfactorily during short duty cycle but the transistors fail when the pulse width is increased to a "full-on" condition. The current during this time approaches or exceeds the continuous rating and the internal connecting wires or clips within the transistor or the external transistor pins may fuse open.

The Ⓦ D60T transistor uses the compression bonding Ⓦ CBE technique of internal connections to eliminate bonding wire shortcomings. A molybdenum pad covers the complete emitter contact area and a heavy copper pole-piece directly contacts this molybdenum pad for current contact. This type of construction completely eliminates emitter wires, wire bonds or other less desirable contacting methods. With this heavy emitter contact system the high current gain, i.e., the gain at points much

higher than the gain rating, is limited only by the transistor gain and not by the internal construction. See figure 1.

Gain Limitations and "Surge" Capability

The D60T transistor is rated at a gain of 10

at 40, 50, or 60 amperes and has a typical gain of 5 at 90 amperes. This high current capability eliminates the need for paralleling transistors in many applications. When paralleling is necessary, it can be accomplished without the problems of fusing and without the need of matching $V_{CE(sat)}$. The

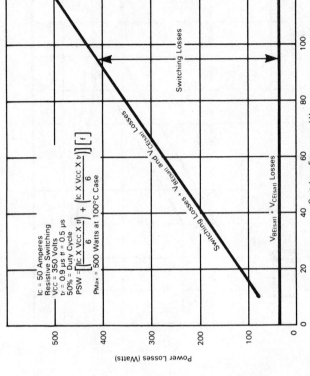

Figure 3: Power Loss vs. Switching Frequency, D60T, Resistive Load

positive temperature coefficient, where the gain decreases and saturation voltage increases as the temperature increases brings one transistor out of saturation forcing the other transistor of a paralleled pair to carry more current. This temperature modulation of the current sharing is possible only in transistors that have high current capability (will not fuse open) and high power capability. Of course, other characteristics such as switching times may have to be matched depending on the circuit and the application.

When the D60T is used at its normal gain rated current of 50 amperes, the gain of 5 at 90 amperes offers the advantage of overload capability. Circuits can be designed to use this added high current capability at 90 amperes as a "surge" safety margin.

Power Gain

Many designers object to driving a transistor with high base currents. However, when the power gain is calculated, a different perspective is achieved. For instance, a base current of 6 amperes is necessary for saturation at a collector current of 50 amperes for the D60T transistor. But when switching these same 50 amperes with a collector voltage of 350 volts, the total peak power switched is 17,500 watts. The base power is only 9 watts, i.e., 1.5 volts x 6 amperes. The current gain is only 8 but the power gain is approximately 2,000.

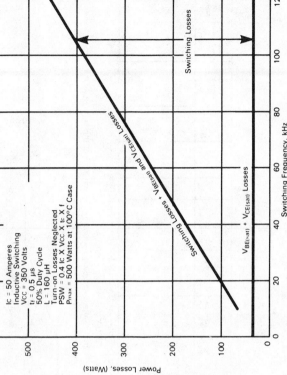

Ic = 50 Amperes
Inductive Switching
VCC = 350 Volts
tf = 0.5 μs
50% Duty Cycle
L = 160 μH
Turn-on Losses Neglected
PSW = 0.4 Ic X VCC X tc X f
Pmax = 500 Watts at 100°C Case

Figure 4: Power Losses vs. Switching Frequency, D60T, Inductive Load

Non-Saturating Circuit

Storage time in transistors is often detrimental in switching circuits. Storage times vary among transistors of the same type and these variations cause difficulties in designing circuits. One of these difficulties is "switch-through", a condition that exists when one transistor turns on while another transistor is in the storage and turn-off mode causing a short circuit across the power supply. Also, one transistor of a parallel pair can turn off before the other due to storage time differences causing a burden on the last transistor to turn off. Storage time can also limit the maximum frequency of operation. Therefore, reducing the storage time by using a non-saturating circuit (Baker's Clamp) as shown in Figure 2 is advisable.

The graph shown in Figure 2 gives the variation of the storage time with a change of load. Diode D1 must be a fast recovery diode and the balance of the diodes are standard recovery time units. This circuit increases the VCE(sat) to about 1.5 volts from the typical 1 volt VCE(sat) of a D60T transistor. This increase in voltage results in a higher power dissipation but this is a small penalty to pay for the added advantages. This non-saturating circuit permits the operation of this transistor at frequencies above 100 kHz. This circuit also reduces the switch-through in standard bridge circuits as well as reducing variations in turn-off times in parallel operation of transistors by reducing the storage times.

Power Dissipation and Frequency of Operation

At low switching frequencies the saturation voltage contributes significant losses but as switching frequencies increase, the switching losses dominate. The D60T transistor is rated at 825 watts power dissipation at 25°C case or 625 watts at 100°C case temperature. With a resistive load, operating at full rated current of 50 amperes and switching 350 volts, the transistor is limited to about 120 kHz as shown in Figure 3.

When switching an inductive load, where the transistor turns on before the current rises, the transistor does not reach the power limitation of 500 watts until about 130 kHz. This inductive load switching is shown in Figure 4. Many of the smaller transistors would have to be paralleled to obtain the power dissipation required to operate at high frequencies. Even when operating at 10 to 20 kHz, the D60T transistor would be preferred because the added power dissipation capability results in lower junction temperatures and greater reliability.

Inductive Turn-Off Capability

All D60T transistors are tested in a clamped inductive circuit shown in Figure 5.

Type Number	Device Package	$V_{CEO(SUS)}$ Volts	I_C x V_{CC} (1) watts (peak)	h_{FE}	at I_C Amperes	h_{FE} x I_C	Power Dissipation 100°C watts	(2) Inductive Switching (100°C) μsec	Figure of Merit (3)
D60T	Stud	400	17,500	10	50	500	525	0.7	35.7
MJ10016	TO-3	500	7,000	20	20	400	143	1.0	5.72
RCA9113B	TO-3	400	1,750	15	5	75	100	1.8	4.16
XGSR15040 (GSI)	TO-3	400	5,250	10	15	150	75	0.5	2.25
2N6547	TO-3	400	3,500	6	10	60	100	1.5	0.4
PT 3523	TO-63	400	17,500	10	50	500	200	1.0(4)	1.0

Table I—Figure of Merit for Inductive Switching

(1) V_{CC} = 350 Volts for all transistors.

(2) t_c is the clamped inductive turn-off time from 10% of the voltage as it begins to rise to the point where the current falls to 10% of its value.

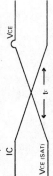

(3) $$\text{Figure of Merit} = \frac{h_{FE}}{10} \times \frac{I_C}{10} \times \frac{\text{Power Dissipation}}{100} \times \frac{10^6}{t_c}$$

(4) Inductive switching not given, t_c is approximated where t_c = 2 X t_f (resistive)

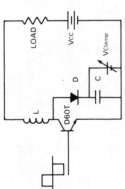

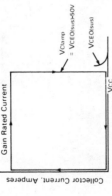

Figure 5: Clamped Inductive Switching Circuit and Load Line

The capacitor is charged to the clamp voltage and then the transistor is switched to the transistor's rated current at 100 pulses per second. The capacitor in the clamp is charged to the transistor $V_{CE(sus)}$ voltage minus 50 volts. This test is assurance that every transistor is capable of switching clamped inductive loads traversing the load line shown in Figure 5.

Figure of Merit

A method of comparing the ratings of transistors in high power switching applications is often needed. Usually the collector cur-rent times the gain (h_{FE} X I_C product) is used. A more useful method would include the inductive turn-off time (t_c) and the power dissipation. The following formula is proposed:

$$\text{Figure of Merit} = \frac{h_{FE}}{10} \times \frac{I_C}{10} \times \frac{\text{Power Dissipation}}{10} \times \frac{10^6}{t_c}$$

Table I shows that the Figure of Merit formula gives better results for comparison than the h_{FE} X I_C number. *Caution must be used when comparing transistors so that the parameters are at the same conditions.* All power dissipations and switching times in Table I were compared at 100°C. The h_{FE} is for the I_C given.

The Figure of Merit indicates that the D60T is 3.5 times better than the nearest listed competition. The D60T transistor has a figure of merit of 35 and the next best transistor has a figure of merit of 10. The D60T and the PT3523 transistors have identical h_{FE} X I_C product and also both transistors are capable of switching the same power of 50 amperes times 350 volts or 17,500 watts. But the PT3523 does not have the same figure of merit because the D60T has a much higher power dissipation capability. The figure of merit as proposed here is capable of a more accurate distinction of parameter differences.

Transistor Paralleling

Before the technical difficulties of paralleling are discussed, it should be pointed out that all bipolar transistors may easily be paralleled when operated in the current source mode. To achieve this type of operation, emitter resistance is added to each device externally, causing each transistor to operate as a current source, in which collector current is nearly independent of collector voltage. Base resistance is also added to each transistor to eliminate the possibility of one transistor taking all the base current. For this type of operation, any number of devices may be paralleled without difficulty if safe operating area constraints are observed.

Many power circuits could be realized by combining small, low-level transistors in parallel. It is, however, more difficult to parallel switching type devices and such circuits must be approached cautiously. If emitter ballasting is used, current sharing should be monitored by viewing the emitter current across the emitter ballast resistors. Both dynamic and static current balance must be considered.

Dynamic current sharing is current sharing during the switching times when collector current in parallel transistors is rising to full load levels or is falling from full load levels. Static current sharing is when paralleled transistors are in the fully saturated, stable conduction period. Production checks of switching transistors in parallel would require satisfactory levels of static and dynamic collector current balance. For good "worst case" current sharing, balance checks should be done at maximum collector junction temperature. These checks are time consuming and expensive.

The parameter related to static current balancing is $V_{CE(sat)}$ which is the small voltage across the switching transistor when it is "on". By using the classic equivalent circuit for the "on" transistor, it is relatively simple to develop equations for a group of (n) transistors showing percent current unbalance[1].

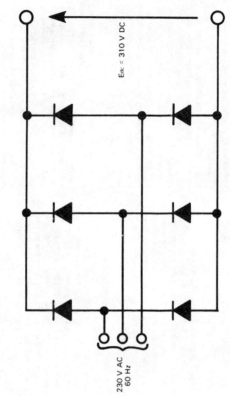

$E_{dc} = 310 \text{ V DC}$

230 V AC
60 Hz

Figure 6: Direct Line Rectification

The indicated reference shows that for ±5% $V_{CE(sat)}$ matching, a collector current unbalance of ±12% is achieved. This means that if transistors with 10 amperes maximum currents are paralleled, the nominal current capacity of the units in parallel can only be considered to be 8.85 amperes. With this matching, current can then increase 12% to 10 amperes, or decrease 12% to 7.9 amperes without exceeding transistor rating. Note that the D60T transistor is gain rated at 40, 50 or 60 amperes but the maximum continuous current is 100 amperes and the maximum pulsed current is 200 amperes.

Some compensating effect exists because of gain-temperature variation and its effect on $V_{CE(sat)}$. $V_{CE(sat)}$ increases with current and temperature. If two transistors in a parallel group differed in current level, a positive temperature coefficient of $V_{CE(sat)}$ would tend to restore current balance and stability.

If $V_{CE(sat)}$ matching can be obtained, it would appear that paralleling switching transistors has been at least 50% solved. The dynamic problems associated with turn-off, however, completely overshadow the static difficulties. The two most important factors of consideration are:

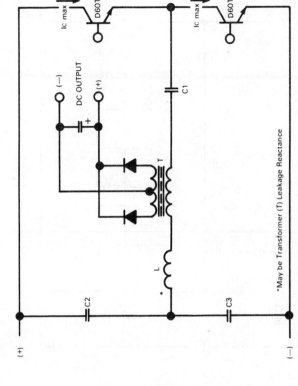

Figure 7: Half Bridge 7.5 kVA DC to DC Converter

*May be Transformer (T) Leakage Reactance

1. Imbalance in transistor turn-off time results in progressive transfer of load current to the last unit to turn off.

2. High speed switching transistors may fail
   due to second breakdown during turn-off
   especially when turning off inductive
   loads.

Transistor manufacturers are now more
completely defining inductive turn-off
capability or "reverse bias SOA". This added
definition will make paralleling more realistic.

Turn-off time reduction by reducing the
storage time using a non-saturating circuit,
and also, the complete characterization of
the inductive load line with 100% testing,
both described earlier, minimizes many of
the problems of paralleling transistors. But
at the present "no model has been advanced which accurately predicts the entire
safe-operating area (SOA) as a function of
quantities which are used in device design[4]"
and those models which describe turn-off
second breakdown are sketchy and ill defined[5], therefore the direct paralleling of
switching transistors must be approached
with the awareness of the hazards of turn-
off. Some of the references already mentioned do present examples of parallel
switching transistors, and paralleling has
been done, but certainly with careful consideration of the difficulties.

Paralleled switching transistors require a
tight characterization of parameters. The
dynamic parameters such as turn-on and
turn-off time must be uniform for safe
operation. This often requires laborious
testing of the switching times but sometimes this is insufficient at higher tempera-

tures, especially where storage time is critical, since it varies with gain and gain varies
with temperature. One solution is the use of
the non-saturating circuit described previously to reduce storage time.

Since paralleling of transistors has many
pitfalls, a single transistor that eliminates
the need for paralleling is preferred. It also
follows, that when paralleling is required,
fewer of the large area, high current transistors will be needed for equal power output making the paralleling task easier.

Power Supply Paralleling

Because it is difficult to parallel switching
transistors, power supplies determined by
the size of the highest capability switching
transistors, are frequently put in parallel.
The result is safe and sure, but not always
the least costly. Duplication of logic level
components, drive circuitry, transformers,
and packaging are some of the items which
add this extra cost. The Westinghouse
D60T will offer relief to power supply
builders who are required to parallel for

greater power. A full bridge could offer 15 kW of output power using four D60T transistors. When operated at 20 kHz, this full bridge inverter output could be rectified to produce 5 volts dc, at the 3000 ampere level. Savings in other areas will similarly be possible using this new and advanced power transistor.

Direct Operation From 230V AC, 60 Hz

Power conditioners using the D60T transistor may use dc power directly derived from the 230V ac, three phase, 60 Hz ac line. The D60T transistor is suitable for use with direct line rectified 60 Hz power. Figure 6 illustrates direct line rectifications.

The voltage resulting from direct rectification of 60 Hz, 230 volts ac line is:

$$E_{dc} = \frac{3\sqrt{2}}{\pi} E_{rms}$$

$$= 310 \text{ Volts dc (for } E_{rms} = 230 \text{ V}_{AC})$$

The sustaining voltage of the D60T switching transistor is 400 to 500 volts. This voltage is sufficient to switch clamped inductive loads directly from rectified 60 Hz, 230 Vac power with the normal safety margins that are expected for good design practice. Transient protection such as clamping or load line shaping must always be used to protect any high voltage transistor.

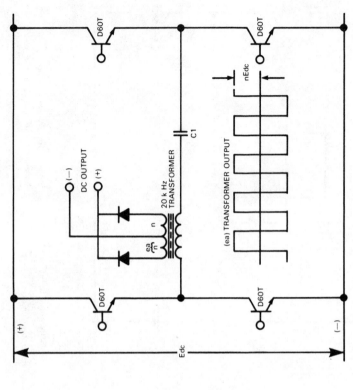

Figure 8: Full Bridge 15 kVA DC to DC Converter

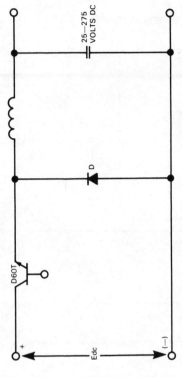

Figure 9: 14 kW DC to DC Chopper Circuit

DC-to-DC Converter—Half Bridge

The rectified dc output from Figure 6 may be fed to an inverter operated at 20 kHz to 100 kHz, with an output rectifier as shown in Figure 7.

For this two-transistor half-bridge circuit, the output power excluding efficiencies, is

$$P_{max} = \frac{E_{dc} \times I_{C \, max}}{2}$$

and the D60T transistor operating from a rectified 230V ac line, is

$$P_{max} = \frac{310 \text{ Volts} \times 50 \text{ Amps}}{2} = 7.5 \text{ kW}$$

The voltage level of the output power is determined by the turns ratio of the output transformer. A wide range of output dc levels is possible, and in practice, fast recovery rectifiers are required for 20-100 kHz rectification. For an intermediate range dc output voltage on 50-350V dc, Westinghouse type R302 and R402 fast recovery rectifiers are suitable as rectification elements. At low voltages such as 5 volts dc, Schottky type rectifiers may be preferred.

The combination of Figures 6 and 7 offers dc-to-dc conversion in a compact size. The elimination of a 60 Hz power transformer makes a large reduction in the weight and volume. Power line isolation is also achieved with the 20-100 kHz, ferrite-type transformer shown in Figure 7.

DC-to-DC Converter—Full Bridge (See Figure 8)

For a full bridge using four transistors as shown, the output power, excluding efficiencies is:

$$P_{max} = E_{dc} \times I_{max}$$

and for D60T transistor operating from rectified 230V ac.

$$P_{max} = 310 \text{ Volts} \times 50 \text{ Amps} = 15.5 \text{ kW}$$

Whether using the half bridge of Figure 7, or the full bridge of Figure 8, each transistor is required to block a maximum voltage of E_{dc} and to conduct a maximum current of 50 amperes for the rated power outputs specified. One problem with the half or full bridge is the unbalances in the switching transistor conducting drops. These drops may cause circulation of a dc current which saturates the output transformer. As a result of this saturation of the transformer, transistor collector current(s) can become excessive, beyond SOA limits, causing transistor failure. The traditional solution of this problem was to gap the output transformer preventing saturation from the anticipated unbalanced dc current. Another solution is to place a dc blocking capacitor (C1) in series with the output transformer as illustrated in Figure 7 and Figure 8. The capacitor blocks any dc circulating current due to transistor conducting unbalance, and prevents transformer saturation.

DC-to-DC Chopper

The single switching regulator of Figure 9 can provide an output voltage of 25-275 volts dc from a rectified source of $E_{dc} = 310$ Volts dc. This can be supplied over the complete voltage range up to load currents of 50 amperes. The power available at 275 volts dc output is 14 kW.

This circuit may use the D60T at chopping rates of 10-20 kHz or higher to give a dc output voltage at similar levels to those available with the high frequency rectification scheme of Figures 6 and 7. The recirculating diode (D) of Figure 9 must be a fast recovery type.

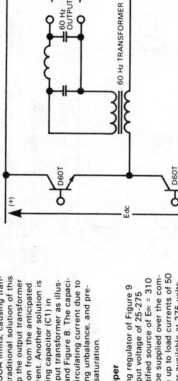

Figure 10: DC-to-AC Fundamental Bridge Inverter

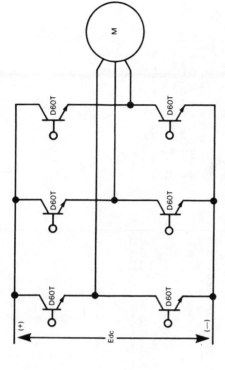

Figure 11: High Performance AC Motor Drive

DC-to-AC Inverter

The full bridge circuit can also be used as a DC-to-AC inverter. A fundamental bridge inverter is shown in Figure 10.

If the nominal peak voltage input to this bridge (E_{dc}) is considered to be 310V dc, and the peak current input is 50 amperes dc, then the maximum output power at the fundamental frequency, excluding efficiencies is:

$$P_{max} = E_{in\ (RMS)} \times I_{in\ (RMS)}$$

$$= .707\ (130) \times .707\ (50)$$

$$P_{max} = 7.75\ kW$$

If the turns ratio of the output transformer is adjusted to produce a nominal 120V ac at the output, then rated output current would be:

$$I_{rated(out)} = I_{in} \times \frac{N_p}{N_s} = I\ \frac{E_p}{E_s}$$

$$I_{rated\ (out)} = .707\ (50) \times \frac{220}{120} = 65\ Amperes$$

In the inverter illustrated, as well as the dc-to-dc converters mentioned previously, numerous control schemes are available which offer a variety of performance levels. These high performance schemes offer superior response and efficiency using the Westinghouse D60T transistor.

AC Motor Drive

A transistorized ac motor drive formed by inverting dc voltage is shown in Figure 11.

Assuming that the inverter will produce 230 volts ac and the peak switchable current is 50 amperes per leg, the output power is:

$$P_{max} = (\sqrt{3})\ (230) \times (50 \times .707) \times Eff.$$

Assuming a 2.5% transistor loss:

$$P_{max} = 14\ kW \times .975 = 13.6\ kW$$

For an 80% efficient motor:

$$Horsepower = \frac{13.6\ kW \times .8\ Eff.}{746\ W/hp}$$

$$Horsepower = 14.5$$

This range is effectively handled by thyristor inverters. However, if the motor to be controlled requires a wide servomechanism bandwidth, such as for precision machine tool applications, then a transistorized motor drive such as shown in Figure 7 has definite performance advantages. As an example, this circuit could easily switch at 5 kHz rates to achieve a 200 Hz servo bandwidth while a comparable thyristor circuit would be operating near the limit of its capability.

Summary

The circuits shown here are the basic circuits that could be used with this transistor.

Variations such as the tuned output full bridge inverter can be designed as well as many other circuits. This transistor has the power dissipation and frequency capability at high voltages and currents making possible operation in areas previously untouched.

The discussion of some of the special characteristics is only lightly covered in this application data sheet. Expert applications assistance as well as data waveform generation for special applications is available from Westinghouse. Please contact Westinghouse should you have a potential D60T application.

Note: The Westinghouse D60T data sheet should be used in conjunction with this application data sheet.

References

1. "Parallel Operation of High Voltage Power Transistors for Motor Speed Control", Arthur P. Connolly, Mariano C. Felix, 1975 IAS Meeting, pp. 467-476.
2. "Mechanisms of Secondary Breakdown in Power Transistors", H. R. Hawkins, Delco Electronics Application Note, May 12, 1977.
3. "Collector Change Dynamics and Second Breakdown of Power Transistor", P.L. Hower, IEEE Power Electronics Specialists Conference, pp. 149-153.
4. "Quantitive Models for Second Breakdown in Transistors", P.L. Hower.
5. "Switching Stress Reduction in Power Transistor Converters", Tore Marvin Underland, 1976 IAS Meeting, pp. 383-392.

Switch 400 volts and 50 amperes up to 100 kHz with the award winning D60T POWER TRANSISTOR

TRUE HIGH POWER CAPABILITY...

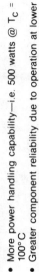

D60T
(23mm diameter)

Competitor A
(12mm square)

Competitor B
(14mm diameter)

Westinghouse's D60T with up to 2.5 times more silicon area provides:

- More power handling capability—i.e. 500 watts @ T_C = 100°C
- Greater component reliability due to operation at lower current density levels

REAL DESIGN BENEFITS...

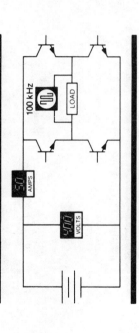

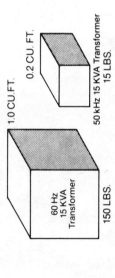

100 kHz

LOAD

AMPS

VOLTS

1.0 CU.FT.

60 Hz
15 KVA
Transformer

150 LBS.

0.2 CU. FT.

50 kHz 15 KVA Transformer

15 LBS.

Multi-Kilowatt, Multi-Kilohertz Operation
Now Possible with Class D and E Switching Circuits

A POWER TRANSISTOR DESIGNED TO LAST...

(Not to scale)

D60T
Massive compression bonded copper emitter contact

Competitor A (Modified TO-3)
Small diameter, point contact wire bonds

Competitor B (TO-63)
thin copper emitter solder contact

Westinghouse's D60T compression bonded encapsulation (CBE) construction with a massive emitter rod contact provides:

- High current overload capability (no bonding wire current limit)
- Better power cycling capability (no solder fatigue)

GET THE FULL STORY...

If you are designing a high power, high frequency switching circuit, or are now paralleling smaller transistors in a high power circuit, consider the Westinghouse D60T:

- Voltage ratings up to 500 volts (sustaining)
- Gain rated at 40 and 50 amperes
- 100 ampere continuous/200 ampere peak current capability
- Inductive switching time turn-off of 500 nanoseconds at 40 amperes

Contact Westinghouse for data sheets, application notes, pricing, and expert application assistance. Westinghouse Electric Corporation, Semiconductor Division, Youngwood, PA 15697. Telephone (412) 925-7272.

- Smaller, lightweight magnetics
- Less filtering capacitance
- Pulse width modulation
- Smaller, lightweight circuits
- Lower parts count/greater reliability
- More energy efficient
- Lower overall systems cost

APPLICATION OPPORTUNITIES...

The D60T power switching transistor has opened the door to previously impractical applications and can improve the performance, economy, and efficiency of many existing applications.

Existing Circuits

- Converters and inverters at 20 kHz
- Switching regulators
- SCR replacement in high frequency choppers
- DC motor drives (PWM)
- Class B amplifiers (linear)

New Application Areas

- Converters and inverters up to 100 kHz
- Switching regulators >5 kilowatts
- Induction heaters and VLF transmitters ≥50 kHz
- AC motor drives and frequency changers
- Class D and E amplifiers (switching mode)

APPENDIX IV
GE C434/C435 SCR

ELECTRONIC
INNOVATIONS
IN ACTION
SEMICONDUCTORS

C434/C435

AMPLIFYING GATE

HIGH SPEED
Silicon Controlled Rectifier
600 Volts (500-700) Amps RMS

The General Electric C434 and C435 Silicon Controlled Rectifiers are designed for power switching at high frequencies. These are all-diffused Press Pak devices employing the field-proven amplifying gate in a new, low profile, low thermal resistant Press Pak housing.

FEATURES:

- Fully characterized for operation in inverter and chopper applications.
- Low thermal resistance (.04°C/W Junction-to-Case).
- High di/dt ratings.
- High dv/dt capability with selections available.
- Rugged hermetic glazed ceramic package having 0.5″ creepage path.

IMPORTANT: Mounting instructions on the last page of this specification **must** be followed.

HIGH FREQUENCY CURRENT RATINGS

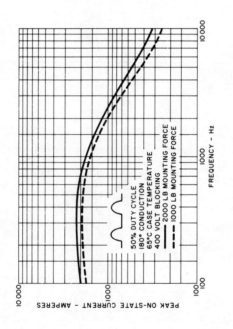

50% DUTY CYCLE
180° CONDUCTION
65°C CASE TEMPERATURE
400 VOLT BLOCKING
—— 2000 LB MOUNTING FORCE
– – – 1000 LB MOUNTING FORCE

PEAK ON-STATE CURRENT – AMPERES

FREQUENCY – Hz

Equipment designers can use the C434 and C435 SCR's in demanding applications, such as:

- Choppers
- Inverters
- Electric Vehicles

- Regulated Power Supplies
- Cycloconverters

- DC to DC Converters
- High Frequency Lighting

MAXIMUM ALLOWABLE RATINGS

TYPES	REPETITIVE PEAK OFF-STATE VOLTAGE, V_{DRM}[1] $T_J = -40°C$ to $+125°C$	REPETITIVE PEAK REVERSE VOLTAGE, V_{RRM}[1] $T_J = -40°C$ to $+125°C$	NON-REPETITIVE PEAK REVERSE VOLTAGE, V_{RSM}[1] $T_J = 125°C$
C434/C435A	100 Volts	100 Volts	150 Volts
C434/C435B	200	200	300
C434/C435C	300	300	400
C434/C435D	400	400	500
C434/C435E	500	500	600
C434/C435M	600	600	720
C434/C435S	700	700	840

[1] Half sinewave waveform, 10 ms. maximum pulse width.

Peak One Cycle Surge (Non-Repetitive) On-State Current, I_{TSM} (60 Hz) – (2000 Lb. Mounting) . . . . 8000 Amperes
Peak One Cycle Surge (Non-Repetitive) On-State Current, I_{TSM} (60 Hz) – (800 Lb. Mounting) . . . . 7500 Amperes
I^2t (for fusing) for times $\geqslant$ 1.5 milliseconds – (2000 Lb.) . . . . 100,000 (RMS Ampere)^2 Seconds
I^2t (for fusing) for times $\geqslant$ 1.5 milliseconds – (800 Lb.) . . . . 93,400 (RMS Ampere)^2 Seconds
I^2t (for fusing) for times $\geqslant$ 8.3 milliseconds – (2000 Lb.) . . . . 250,000 (RMS Ampere)^2 Seconds
I^2t (for fusing) for times $\geqslant$ 8.3 milliseconds – (800 Lb.) . . . . 233,000 (RMS Ampere)^2 Seconds
Critical Rate-of-Rise of On-State Current, Non-Repetitive. . . . . 800 A/μs †
Critical Rate-of-Rise of On-State Current, Repetitive. . . . . 500 A/μs †
Average Gate Power Dissipation, $P_{G(AV)}$ . . . . 2 Watts
Storage Temperature, T_{stg} . . . . -40°C to +150°C
Operating Temperature, T_J . . . . -40°C to +125°C
Mounting Force Required . . . . 800 Lbs. Min. (3.6 Kn)
2500 Lbs. Max. (11.1 Kn)

†di/dt ratings established in accordance with EIA-NEMA Standard RS-397, Section 5.2.2.6 for conditions of max rated V_{DRM}; 20 volts, 20 ohms gate trigger source with 0.5 μs short trigger current rise time.

CHARACTERISTICS

TEST	SYMBOL	MIN.	TYP.	MAX.	UNITS	TEST CONDITION
Repetitive Off-State and Peak Reverse Current	I_{RRM} and I_{DRM}	–	5	15	mA	$T_J = +25°C$, $V_{DRM} = V_{RRM}$
Repetitive Off-State and Peak Reverse Current	I_{RRM} and I_{DRM}	–	20	45	mA	$T_J = 125°C$, $V_{DRM} = V_{RRM}$
Thermal Resistance (DC)	$R_{\theta JC}$	–	–	.04	°C/Watt	Junction-to-Case, Double-Side Cooling, 2000 lbs.
		–	–	.05		Junction-to-Case, Double-Side Cooling, 800 lbs.
Critical Rate-of-Rise of Forward Blocking Voltage (Higher values may cause device switching)	dv/dt	200	500	–	V/μsec	$T_J = +125°C$, Gate Open. V_{DRM} = Rated Linear or Exponential Rising Waveform Exponential dv/dt = $\dfrac{V_{DRM}}{\tau}$ (.632)
						Higher minimum dv/dt selections available – consult factory.
Holding Current	I_H	–	40	1000	mAdc	$T_C = +25°C$, Anode Supply = 24 Vdc, Initial On-State Current = 10 Amps.
DC Gate Trigger Current	I_{GT}	–	70	200	mAdc	$T_C = +25°C$, $V_D = 10$ Vdc, $R_L = 1$ Ohm
		–	100	400		$T_C = -40°C$, $V_D = 10$ Vdc, $R_L = 1$ Ohm
		–	25	150		$T_C = +125°C$, $V_D = 10$ Vdc, $R_L = 1$ Ohm
DC Gate Trigger Voltage	V_{GT}	–	3	5	Vdc	$T_C = -40°C$ to $+25°C$, $V_D = 10$ Vdc, $R_L = 1$ Ohm
		–	1.50	3.0		$T_C = +25°C$ to $+125°C$, $V_D = 10$ Vdc, $R_L = 1$ Ohm
		0.15	–	–		$T_C = +125°C$, V_{DRM}, $R_L = 500$ Ohms
Peak On-State Voltage	V_{TM}	–	2.3	2.5	Volts	$T_C = +25°C$, $I_{TM} = 3000$ Amps. Peak Duty Cycle ⩽ .01%. Pulse Width = 3.0 ms.

Parameter	Symbol	Min	Typ	Max	Units	Test Conditions
Turn-On Delay Time	t_d	—	0.5	—	μsec	$T_C = +25°C$, $I_{TM} = 50$ Adc, V_{DRM}. Gate Supply: 20 Volt Open Circuit, 20 Ohms, 0.1 μsec. max. rise time. ††, †††
Conventional Circuit Commutated Turn-Off Time (with Reverse Voltage)	t_q				μsec	(1) $T_C = +125°C$ (2) $I_{TM} = 500$ Amps (3) $V_R = 50$ Volts Min. (4) V_{DRM} (Reapplied) (5) Rate-of-Rise of reapplied off-state voltage = 20 V/μsec (linear) (6) Commutation di/dt = 25 Amps/μsec (7) Repetition rate = 1 pps (8) Gate bias during turn-off interval = 0 volts, 100 ohms
C434		—	8	†		
C435		—	12	†		
Conventional Circuit Commutated Turn-Off Time (with Reverse Voltage)	t_q				μsec	(1) $T_C = +125°C$ (2) $I_{TM} = 500$ Amps (3) $V_R = 50$ Volts Min. (4) V_{DRM} (Reapplied) (5) Rate-of-Rise of reapplied off-state voltage = 200 V/μsec (linear) (6) Commutation di/dt = 25 Amps/μsec (7) Repetition rate = 1 pps. (8) Gate bias during turn-off interval = 0 volts, 100 ohms.
C434		—	12	14		
C435		—	17	20		
Conventional Circuit Commutated Turn-Off Time (with Feedback Diode)	$t_{q(diode)}$				μsec	(1) $T_C = +125°C$ (2) $I_{TM} = 500$ Amps (3) $V_R = 1$ Volt (4) V_{DRM} (Reapplied) (5) Rate-of-Rise of reapplied off-state voltage = 200 V/μsec (linear) (6) Commutation di/dt = 25 Amps/μsec (7) Repetition rate = 1 pps. (8) Gate bias during turn-off interval = 0 volts, 100 ohms
C434		—	18	†		
C435		—	25	†		

† Consult factory for specified maximum turn-off time.

†† Delay time may increase significantly as the gate drive approaches the I_{GT} of the Device Under Test.

††† Current risetime as measured with a current probe, or voltage risetime across a non-inductive resistor.

SINEWAVE CURRENT DATA

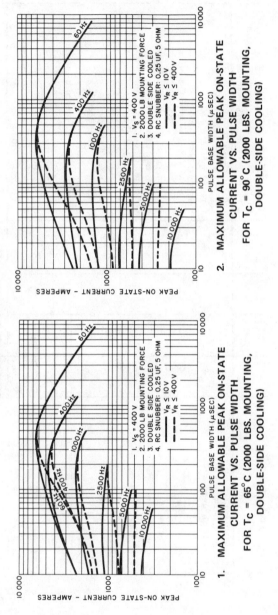

1. **MAXIMUM ALLOWABLE PEAK ON-STATE CURRENT VS. PULSE WIDTH FOR $T_C = 65°C$ (2000 LBS. MOUNTING, DOUBLE-SIDE COOLING)**

2. **MAXIMUM ALLOWABLE PEAK ON-STATE CURRENT VS. PULSE WIDTH FOR $T_C = 90°C$ (2000 LBS. MOUNTING, DOUBLE-SIDE COOLING)**

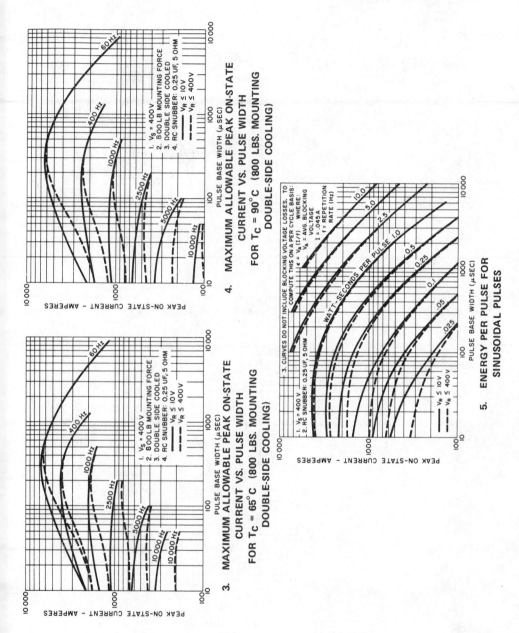

3. MAXIMUM ALLOWABLE PEAK ON-STATE
CURRENT VS. PULSE WIDTH
FOR T_C = 65°C (800 LBS. MOUNTING
DOUBLE-SIDE COOLING)

4. MAXIMUM ALLOWABLE PEAK ON-STATE
CURRENT VS. PULSE WIDTH
FOR T_C = 90°C (800 LBS. MOUNTING
DOUBLE-SIDE COOLING)

5. ENERGY PER PULSE FOR
SINUSOIDAL PULSES

TRAPEZOIDAL WAVE CURRENT DATA

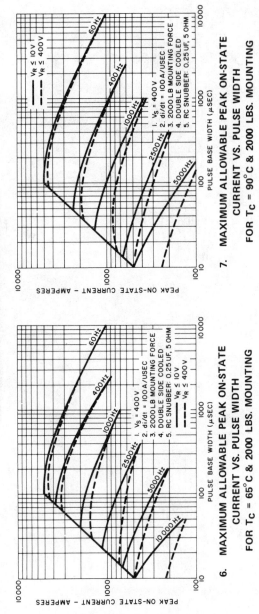

6. MAXIMUM ALLOWABLE PEAK ON-STATE
   CURRENT VS. PULSE WIDTH
   FOR T_C = 65°C & 2000 LBS. MOUNTING

7. MAXIMUM ALLOWABLE PEAK ON-STATE
   CURRENT VS. PULSE WIDTH
   FOR T_C = 90°C & 2000 LBS. MOUNTING

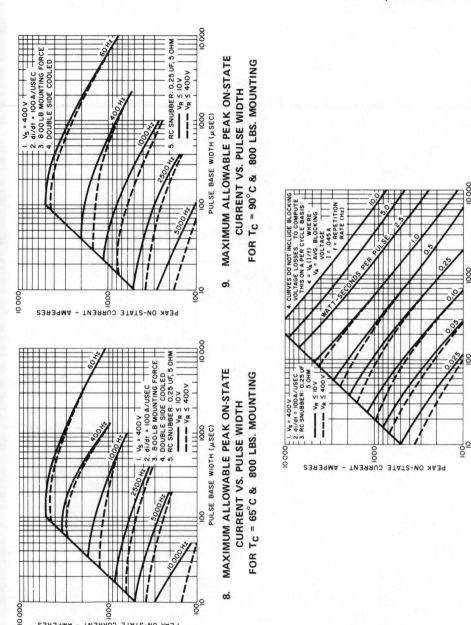

8. MAXIMUM ALLOWABLE PEAK ON-STATE CURRENT VS. PULSE WIDTH FOR T_C = 65°C & 800 LBS. MOUNTING

9. MAXIMUM ALLOWABLE PEAK ON-STATE CURRENT VS. PULSE WIDTH FOR T_C = 90°C & 800 LBS. MOUNTING

10. ENERGY PER PULSE FOR TRAPEZOIDAL CURRENT PULSES

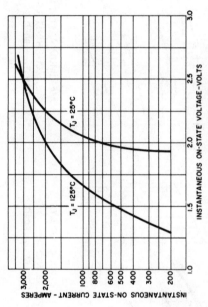

INSTANTANEOUS ON-STATE CURRENT – AMPERES

$T_J = 125°C$

$T_J = 25°C$

INSTANTANEOUS ON-STATE VOLTAGE–VOLTS

11.   MAXIMUM ON-STATE CHARACTERISTICS

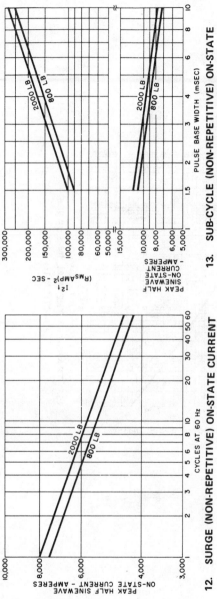

13. SUB-CYCLE (NON-REPETITIVE) ON-STATE CURRENT AND I²t RATING

12. SURGE (NON-REPETITIVE) ON-STATE CURRENT

NOTES:

1. The locus of possible dc trigger points lie outside the boundaries shown at various case temperatures.

2. 20V – 20Ω is the minimum gate source load line when rate of circuit current rise > 100 amp/µs or anode rate of current rise > 200 amps/µs (Tp = 5 µs min., 0.5 µs max. rise time). Maximum, long-term repetitive anode, di/dt = 500 amps/µs with 20V – 20Ω gate source.

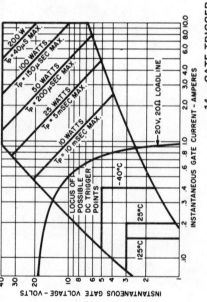

14. GATE TRIGGER CHARACTERISTICS POWER RATINGS

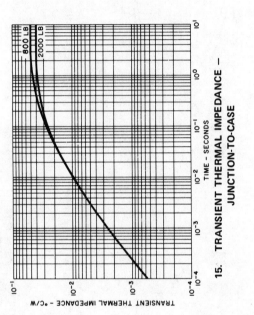

15. TRANSIENT THERMAL IMPEDANCE — JUNCTION-TO-CASE

SYM	INCHES MIN	INCHES MAX	METRIC–MM MIN	METRIC–MM MAX
A	.560	.605	14.22	15.37
B	.985	.995	25.01	25.27
C	1.600	1.650	40.64	41.91
D	.030	–	.76	–
E	.040	–	1.01	–
G	.057	.059	1.44	1.50
H	.186	.191	4.72	4.85
J	.245	.255	6.22	6.48
K	.115	.130	2.92	3.30
L	.064	.070	1.62	1.78
M	–	1.120	–	28.45
N	–	1.585	–	40.26
P	.135	.145	3.42	3.68
Q	.070	.080	1.77	2.01
R	–	.875	–	22.23
S	1.2219	1.2343	31.036	31.351
T	.137	.151	3.47	3.87

OUTLINE DRAWING

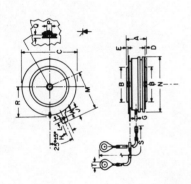

SUGGESTED MOUNTING METHODS FOR PRESS-PAKS TO HEAT DISSIPATORS

When the Press-Pak is assembled to a heatsink in accordance with the following general instructions, a reliable and low thermal resistance interface with result.

1. Check each mating surface for nicks, scratches, flatness and surface finish. The heat dissipator mating surfaces should be flat within .0005 inches and have a surface finish of 63 micro-inches.

2. It is recommended that the heat dissipator be plated with nickel, tin or gold iridite. Bare aluminum or copper surfaces will oxidize in time resulting in excessively high thermal resistance.

3. Sand each surface *lightly* with 600 grit paper just prior to assembly. Clean off and apply silicone oil (GE SF1154 200, centistoke viscosity) or silicone grease (GE G623 or Dow Corning DC 3, 4, 340 or 640). Clean off and apply again as a *thin* film. (A thick film will adversely affect the electrical and thermal resistances.)

4. Assemble with the specified mounting force applied through a self-leveling, swivel connection. The force has to be evenly distributed over the full area. Center holes on both top and bottom of the Press-Pak are for locating purposes only.

INDEX